STERLING

Education

Essential Chemistry

Chemical Thermodynamics

4th edition

Customer Satisfaction Guarantee

Your feedback is important because we strive to provide the highest quality educational materials. Email us comments or suggestions.

info@sterling–prep.com

We reply to emails – check your spam folder

4 3 2 1

ISBN-13: 979-8-8855706-7-1

Sterling Education materials are available at quantity discounts.

Contact info@sterling–prep.com

Sterling Education
6 Liberty Square #11
Boston, MA 02109

Published by Sterling Education

Printed in the U.S.A.

STERLING
Education

From the foundations of chemical reactions to the complex mechanisms of atomic particles, *Essential Chemistry Self-Teaching Guides* are a comprehensive compendium of clearly explained texts to learn and master these multifaceted chemistry topics.

These guides provide a detailed review of the fundamental mechanisms of chemical and physical processes at the atomic level. Develop a better understanding of the electronic structure of elements, principles of chemical bonding, phases of matter, types and mechanisms of chemical reactions, and principles of solutions and acid-base equilibria. Learn about rate processes in chemical reactions, empirical and molecular formulas, enthalpy, entropy, oxidation number, the laws of thermodynamics, and electrochemistry. Reinforce your learning by working through the practice questions and step-by-step solutions.

Created by highly qualified chemistry instructors, researchers, and education specialists, these books empower readers by helping them increase their understanding of general chemistry.

We sincerely hope that these guides are valuable for your learning.

250909akp

Essential Chemistry Self-Teaching Guides

Electronic Structure & Periodic Table

Chemical Bonding

States of Matter & Phase Equilibria

Stoichiometry

Solution Chemistry

Chemical Kinetics & Equilibrium

Acids & Bases

Chemical Thermodynamics

Electrochemistry

Visit our Amazon store

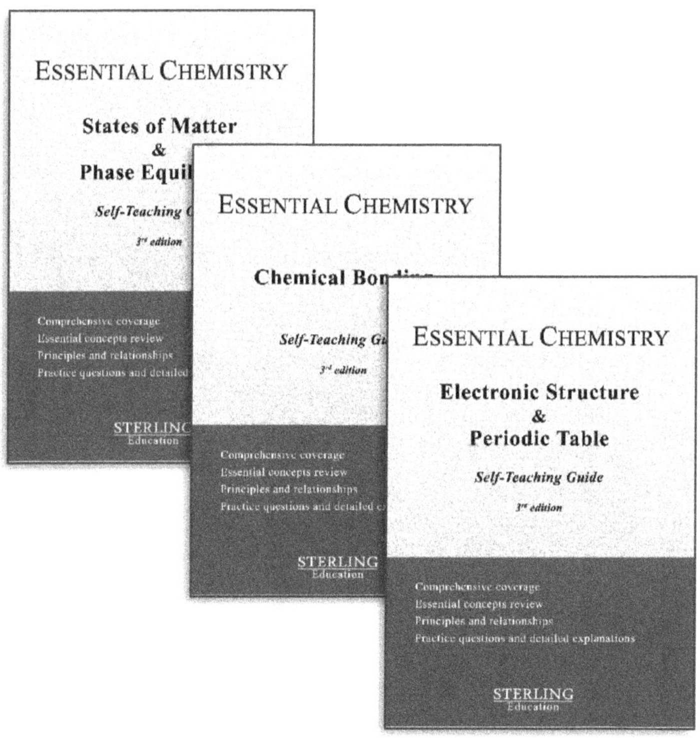

Essential Physics Self-Teaching Guides

Kinematics and Dynamics

Equilibrium and Momentum

Force, Motion and Gravitation

Work and Energy

Fluids and Solids

Waves and Periodic Motion

Light and Optics

Sound

Electrostatics and Electromagnetism

Electric Circuits

Thermal Physics

Atomic and Nuclear Physics

Visit our Amazon store

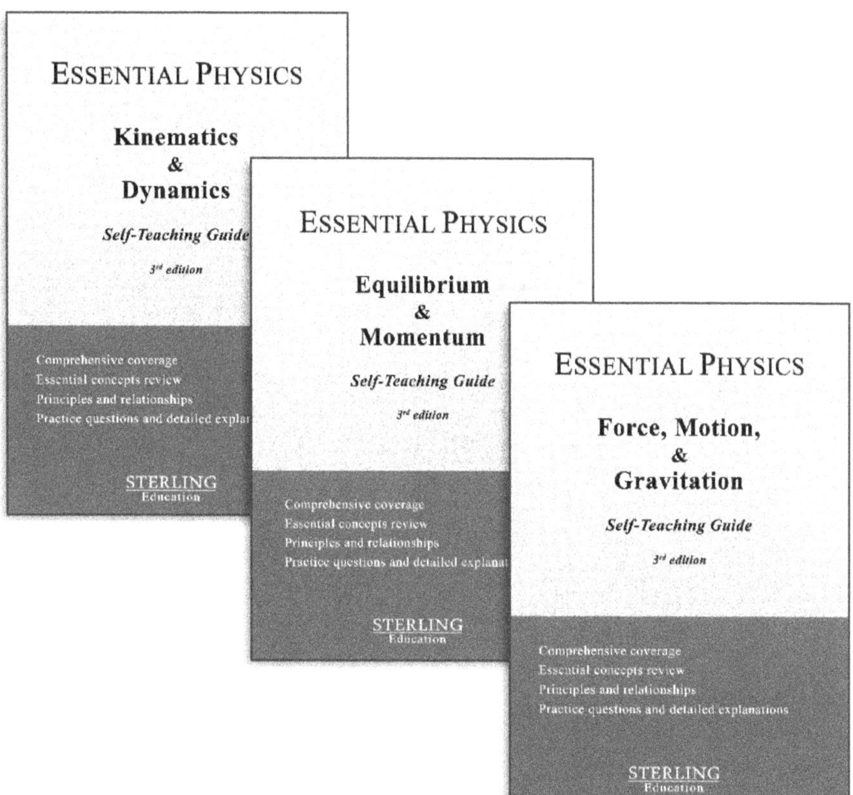

Essential Biology Self-Teaching Guides

Eukaryotic Cell & Cellular Metabolism

Molecular Biology & Genetics

Nervous & Endocrine Systems

Circulatory, Respiratory & Immune Systems

Digestive & Excretory Systems

Muscle, Skeletal & Integumentary Systems

Reproduction & Development

Microbiology

Plants & Photosynthesis

Evolution, Classification & Diversity

Ecology & Population Biology

Visit our Amazon store

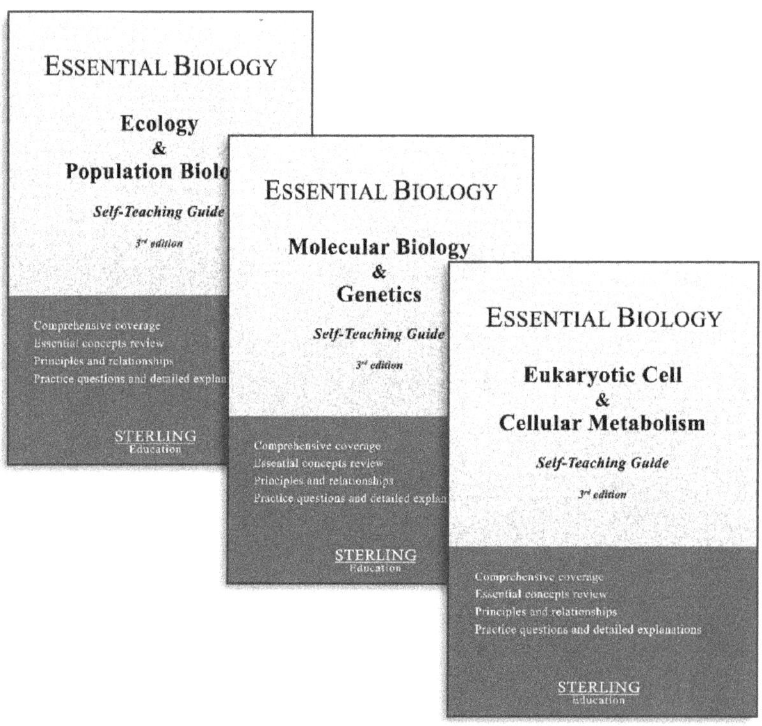

Everything You Always Wanted to Know About…

Chemistry

Physics

Cell & Molecular Biology

Organismal Biology

Human Anatomy & Physiology

Environmental Science

American History

American Law

American Government & Politics

Comparative Government & Politics

World History

European History

Psychology

Sociology

Human Geography

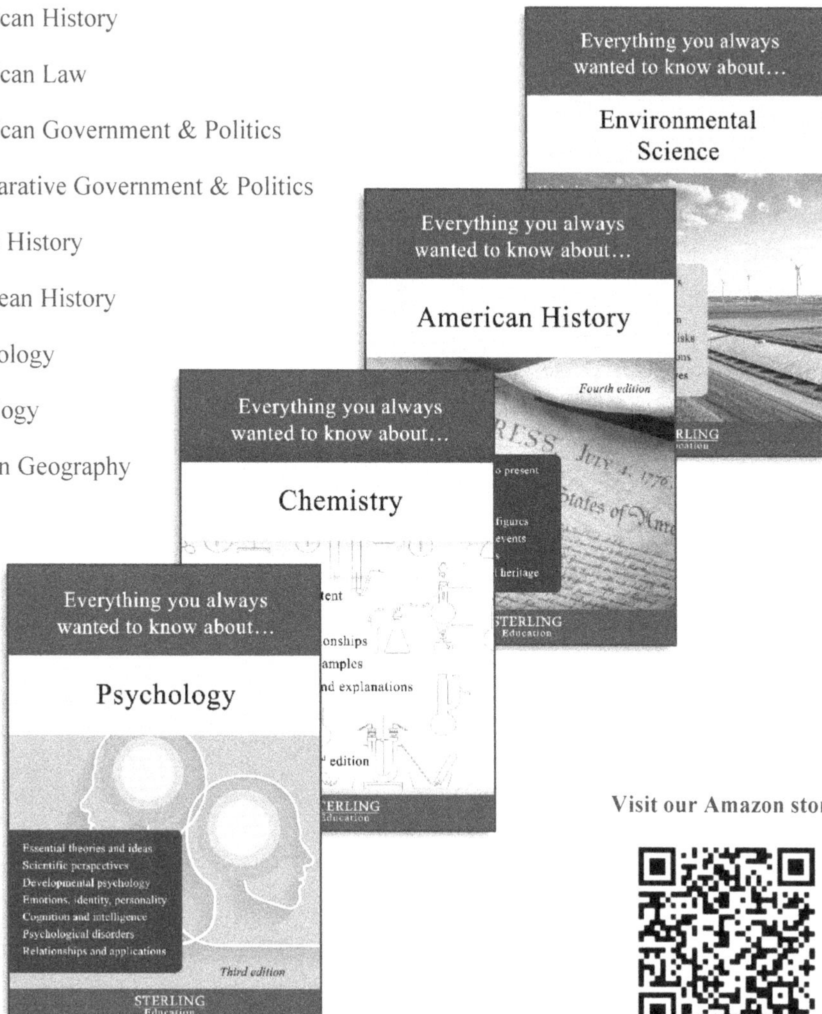

Visit our Amazon store

For online practice resources visit

https://www.sterling-prep.com

If you benefited from this book, please leave a review on Amazon so others can learn from your input. Reviews help us understand our customers' needs and experiences while keeping our commitment to quality.

Table of Contents

Table of Contents (*continued*)

Table of Contents (*continued*)

REVIEW: Chemical Thermodynamics (*continued*)

Table of Contents (*continued*)

Table of Contents (*continued*)

Table of Contents (*continued*)

REVIEW: Chemical Thermodynamics (*continued*)

Table of Contents (*continued*)

Table of Contents (*continued*)

REVIEW

Chemical Thermodynamics

Page intentionally left blank

The Nature of Science

Scientific discovery

Observation is the first step toward scientific discovery. Observations recognize patterns and anomalies in unexplained phenomena.

Hypothesis is advanced as a proposed explanation for the observation.

Evaluating a hypothesis involves experiments and additional observations (i.e., recording and collecting data).

Experimental observations are made while manipulating *independent variables* (e.g., how long the plant is exposed to sunlight per twenty-four hours) with *dependent variables* (data collected regarding height measurements).

Observations generate *data* that:

1) *support the hypothesis* (i.e., different from proves) or

2) *refute the hypothesis*.

Hypothesis are modified by narrowing or expanding their application to explain additional data. More experiments are performed, and the results either *support* (not prove) or *refute* the hypothesis during this iterative process.

Principles are the initially proposed explanations that are specific and apply to a narrow range of phenomena.

Theories build upon valid hypotheses

Theories are proposed to account for the observations when a hypothesis is valid over a range (e.g., sunlight, temperature, or humidity).

Theories are advanced to explain the phenomena and to predict what will happen under other conditions with different variables.

Further observations determine whether the predictions were accurate.

Research (re-search or *again search*) continues as these many observations support hypotheses and theories.

When theories are accurate over various scenarios and experiments, *Laws* are developed based on these repeatedly replicated results.

Laws as established theories

Laws are brief descriptions of how nature behaves in a broad set of circumstances. They are the consolidation of several theories. Laws are the products of multiple theories tested and proven, culminating in a broad claim regarding those predictions.

Laws are widely applicable explanations. Laws with broad and straightforward claims are more robust than those that make claims on a narrower, more specific set of variables and parameters.

The number of Laws describing physical phenomena is minuscule compared to the number of theories and magnitudes, which are less than the number of supported hypotheses.

The number of refuted (unsupported) hypotheses may be several magnitudes larger.

Failed experiments do not typically refer to the physical process where the researcher made a human error during the experiment. A failed experiment means that the *data does not support the hypothesis.* Therefore, after additional trials to validate the "failed" results, the hypothesis must be modified to predict the results of either future experiments or objective observations in the physical world.

Theories, models and laws

Theories emerge from a hypothesis that experimental data has repeatedly supported. Theories are detailed statements that provide testable predictions of the behavior of natural phenomena.

Theories are established to explain observations and then tested (supported or refuted) based on their predictions. Tested theories can articulate the boundaries of the tested hypothesis.

Theories are more specific than *hypotheses* but vaguer and more encompassing than *models*.

Models are constructed to explain the observed behavior if a hypothesis accurately predicts a phenomenon. They are more specific in their application than theories.

Data collected to develop the theories provide the conceptual foundation for constructing models. The model creates mental pictures (e.g., complicated weather patterns displayed as visual pictures for the viewer) of the physical phenomena.

Models should be consistent with scientific theories' predictions and support a comprehensive understanding of a phenomenon. Models allow for predictive outcomes under specific theories (e.g., when the wind speed reaches x, then y will occur).

Multiple models are integrated to explain a portion, or the entire scope, of a *theory*.

Understand the limitations of a model and do not apply it dogmatically unless its applicability is evaluated. A model is an underlying representation of a part of a theory; it is not intended to provide a complete picture of every phenomenon under that theory.

Instead, models provide the fundamental aspects of the theory.

Thermochemistry

Applied thermochemistry

Thermochemistry:

- explains *energy changes*, particularly for energy exchange between the system and its surroundings;
- predicts reactant and product *quantities* during the reaction;
- predicts if a reaction is *spontaneous or non-spontaneous*, favorable, or unfavorable;
- combines concepts of *thermodynamics and chemical bond energy.*

Thermochemistry quantities

Thermochemistry commonly includes calculations of such *quantities* as:

heat capacity (c)

heat of combustion

heat of formation

enthalpy (H)

entropy (S)

Gibbs free energy (G)

calories

Energy changes in chemical reactions

Thermochemistry studies heat energy associated with chemical reactions and physical transformations.

Reactions may *release energy* (i.e., exothermic reactions) or *absorb energy* (i.e., endothermic reactions).

Phase change may *release or absorb energy* (e.g., boiling, melting).

Thermochemistry rests on *two* generalizations:

1. Energy change accompanying any transformation is equal and opposite to the energy change of the reverse process. (Lavoisier and Laplace law proposed in 1780).
2. Hess' Law, proposed in 1840, states that the energy change of any transformation is the same whether the process occurs in one step or multiple steps.

Systems conserve energy

German physicist Julius Robert Mayer (1814-1878) first proposed the *law of the conservation of energy* in 1842, now called the First Law of Thermodynamics.

Conservation of energy law states that *energy is neither created nor destroyed.*

These generalizations were the foundations for *specific heat, latent heat,* and *heat capacity*:

$$\Delta C_p = d\Delta H / dT$$

Integration of the expression $\Delta C_p = d\Delta H / dT$ evaluates the heat of reactions at different temperatures.

Heat and Temperature

Heat as energy

Heat is energy that exists independently of any medium; it is a means of transferring *energy*.

Heat and temperature are often used interchangeably in everyday vernacular.

However, this usage of interchanging heat and temperature is incorrect.

Temperature as kinetic motion

Temperature measures *kinetic energy* (KE) in a specific medium and depends on the material.

Temperature and heat are closely related principles, but *heat is total energy* resulting from *molecular motion.*

In contrast, temperature is the *average* measure of *kinetic energy* in a substance.

Some substances respond drastically to a change in thermal energy; response depends on the material and heat capacity.

Thermal energy

The quantity of thermal energy transmitted from one body to another is measured as heat.

An object does *not* "contain" heat but can have a certain amount of thermal energy.

Heat can be measured in joules (J) and calories (cal).

Microscopically, thermal energy is related to atomic and molecular *vibrations*.

Matter is made of billions and trillions of small molecules; even in solids, these molecules and atoms are never entirely still.

Molecular movement determines the amount of *thermal energy* they possess.

Notes for active learning

Temperature Scales

Fahrenheit and Celsius

Three commonly used temperature scales are *Fahrenheit*, *Celsius* and *Kelvin*.

Daniel Fahrenheit (1686-1736) proposed the classic English system of measuring temperature (i.e., °F) in 1724 by dividing the difference between the boiling and freezing points of water into 180 equal degrees (°F).

Anders Celsius (1701-1744) developed the modern metric system of measuring temperature by dividing the boiling and freezing points of water into 100 equal degrees (°C).

Celsius scale was originally the *centigrade scale*.

In the United States, the temperature unit is *Fahrenheit* (°F).

However, a unit of temperature is the Celsius scale (°C), with water freezing at 0 °C and boiling at 100 °C (instead of 32 °F and 212 °F).

Conversion between Fahrenheit and Celsius:

$$°\mathrm{F} = \frac{9}{5}°\mathrm{C} + 32$$

Kelvin scale

William Thomas, 1st Baron Kelvin (1824-1907), developed the Kelvin (K) as the third temperature scale. Kelvin scale is based on the concept of absolute zero (i.e., 0 K, –273.15 °C, or –459.67 °F) and is the most fundamental scale for measuring temperature in science.

Celsius scale is derived from the Kelvin scale, with 0°C being equal to 273.15 K.

The size of units on the Kelvin scale is the *same* as for degrees on the Celsius scale.

Kelvin is another unit of temperature often used in scientific calculations. This scale is often used for gases, as usually, low temperatures are in question.

Absolute zero is the coldest theoretical temperature any substance can have, equal to 0 K.

0 K is equivalent to –273 °C!

Relative temperature scales

Three thermometers display equal temperatures and three standard temperature units are shown.

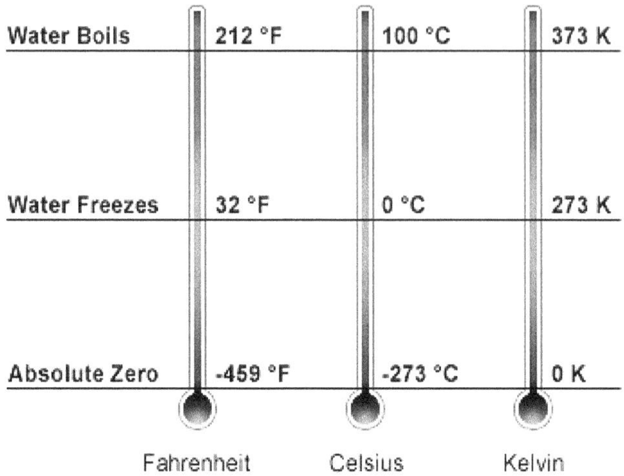

Comparison between the three temperature scales

Converting between scales

Temperature is the average amount of *kinetic energy* of the particles (i.e., molecules and atoms).

Absolute zero is when these particles theoretically have no KE and no movement.

Kelvin and Celsius use the same step in their degree measurements, making calculations easier.

For example, a temperature difference is the same in Kelvin as in Celsius, so when calculating a temperature difference in either of these units, the values do not need to be converted.

If required to convert between Kelvin and Celsius, the conversion can be performed by adding 273 to the Celsius temperature to get Kelvin:

$$K = °C + 273$$

The temperature of absolute zero may never be attained. Scientists have achieved temperatures close to absolute zero but have never fully reached it.

In 2003, the lowest temperature recorded was by a group of scientists from the Massachusetts Institute of Technology (MIT). Researchers successfully cooled sodium (Na) gas to a temperature of half a billionth of a degree above absolute zero, but not to absolute zero itself.

Fahrenheit and Celsius temperature conversions

Convert between Fahrenheit and Celsius by:

$$°F = (1.8 \times °C) + 32$$

For example, what is the temperature in °F for 28 °C?

$$°F = (1.8 \times °C) + 32$$

$$°F = (1.8 \times 28\ °C) + 32$$

$$°F = (50.4) + 32$$

$$°F = 82.4$$

Therefore, 28 °C equals 82.4 °F

For example, what is the temperature in °C for 70 °F?

$$°F = (1.8 \times °C) + 32$$

Rearrange:

$$(1.8 \times °C) = °F - 32$$

$$(1.8 \times °C) = °F - 32$$

$$°C = (°F - 32) / 1.8$$

$$°C = (70 - 32) / 1.8$$

$$°C = 21$$

Therefore, 70 °F equals 21 °C

Kelvin and Celsius temperature conversions

Convert between Kelvin and Celsius:

$$K = °C + 273$$

For example, what is the temperature in °C for 295 K?

$$K = °C + 273$$

Rearrange:

$$°C = K - 273$$

$$°C = 295 - 273$$

$$°C = 22$$

Therefore, 295 K = 22 °C.

Reference values among temperature scales

	K	°C	°F
Absolute zero	0	–273	–460
Freezing point of water / melting point of ice	273	0	32
Room temperature	298	25	77
Body temperature	310	37	99
Boiling point of water / condensation of steam	373	100	212

Heat Capacity and Specific Heat

Heat capacity, specific heat and molar heat capacity

	Heat capacity	Specific heat capacity	Molar heat capacity
Definition	Amount of heat required to raise temperature of a substance by 1 K (1 °C) Ability of a substance to absorb heat	Amount of heat required to raise the temperature of 1 g of a substance by 1 K (1 °C) Amount of heat added (or removed) to change the temperature	Amount of heat required to raise temperature of 1 mole of substance by 1 K (1 °C) At constant pressure unless noted
Equation	$C_H = q / \Delta T$	$C_S = q / m\Delta T$ C_S (specific) equals C_P (constant pressure)	$C_M = q / n\Delta T$
Intrinsicality	Intrinsic property	Non-intrinsic property	
Unit	J / °C	J / g °C or J / Kg K	J / mol °C or J / mol K
Property	Extensive property	Intensive property	Intensive property
Substance quantity	Depends on amount	Independent of amount	
Specificity	Non-specific	Highly specific	
Examples	1 L H_2O = 4182 J/kg °C 100 g Cu = 38.5 J/ °C	H_2O = 1 calorie 1 calorie = 4.18 J/g °C Cu = 0.385 J/g °C	

Heat capacity

Heat capacity (or *thermal capacity*) is a physical property of a substance's ability to hold heat.

Heat capacity (C_H) of a substance is the energy required to raise its temperature by *one degree Celsius*. It is the ratio of absorbed heat energy to the resulting temperature change.

Heat capacity is a product of an *object's mass* (m) and a *substance's specific heat capacity (c)*.

When measuring heat capacity of a substance, a variable must be held constant, as the heat capacity of a system depends on its temperature, pressure, and volume. These are noted with either a c_p or a c_v, respectively.

SI unit for heat capacity is the joule per Kelvin (J/K).

Isobaric and isochoric processes

Gases and liquids are typically measured at constant volume and subjected to pressure for a constant reading.

> *Isobaric process* (c_p) is when *pressure* is held constant, defining the *heat capacity.*
>
> *Isochoric process* (c_v) is when *volume* is held constant.

Heat capacity *vs.* specific heat capacity

Heat capacity and specific heat of substances are similar.

> *Specific heat* (C_S) is the ratio of energy absorbed to the temperature rise per unit of mass of the substance.
>
> *Heat capacity* is the ratio of energy absorbed to the rise in temperature.

Specific heat, c, is a characteristic of the substance and thus remains constant.

Water has a relatively *high* specific heat capacity; 4.2 J of heat energy is required to raise the temperature of one gram of water by 1 °C.

High specific heat of water helps *moderate global temperatures*. Temperature fluctuations during the day and night would be much larger if the water's heat capacity were low. This is evident in a desert where water is scarce. Daytime temperatures are high, yet nighttime temperatures can drop to near freezing.

Specific heat

Specific heat capacity (C_S) is the amount of heat required to raise the temperature of 1 gram of a substance by 1 °C.

Specific heat, c, is a characteristic of the substance and thus remains constant. Metals tend to have *low* specific heat capacities.

Heat capacity and the specific heat of a substance are similar.

Specific heat capacity differs from heat capacity because it relates to the *amount of heat energy* required to raise the temperature of one gram of a substance by one degree Celsius.

> *Specific heat capacity* (or *specific heat*) is the heat capacity of 1 gram of a substance.
>
> *Molar heat capacity* is the heat capacity of 1 mole of a substance.

Mass (g)

Temperature change (°C)

Heat (J) → $$q = m \times C_s \times \Delta T$$

Specific heat capacity (J/g · °C)

Specific heat capacity

Specific heat relates to heat and temperature

Units for specific heat are calories or joules per gram per °C. For example, the specific heat of water is 1 calorie (or 4.18 joules) per gram per Celsius degree.

Specific heat capacity of water is 4,180 joules per kilogram per degree Celsius (J/kg°C); 4,180 J are needed to raise the temperature of 1 kg of water by 1 °C.

Specific heat is related to heat and temperature by:

$$q = mC_S\Delta T$$

where q is the heat flow, m is mass measured in grams, C_S is the specific heat capacity of the substance, and ΔT is the temperature change.

Specific heat (C_S) of a substance, heat (Q), and temperature are related by:

$$Q = mC_S\Delta T$$

where Q is the heat transferred to the material (J), m is the mass of the object being heated (kg), and ΔT is the temperature change (K).

Calculating specific heat

For example, 3,200 J of heat is added to 1.0 kg of water at 10 °C. (Use specific heat of water C_S = 4,190 J/kg·°C)

Rearrange the specific heat equation to solve ΔT:

$$Q = mC_S\Delta T$$

$$\Delta T = \frac{Q}{mc}$$

$$\Delta T = \frac{3{,}200\,J}{1.0\,kg \cdot 4{,}190\,\frac{J}{kg} \cdot °C}$$

$$\Delta T = 0.76\ °C$$

where Q is the heat transferred to the material (J), m is the mass of the object being heated (kg), and ΔT is the temperature change (K).

ΔT represents the temperature change. This value for the change (not the absolute value) must be either added to the initial temperature to determine the final temperature or subtracted from the final temperature to find the initial temperature.

A reaction must be contained within a system to observe all the products.

Specific heats of selected materials

Material	*c*(J/kg·K)
Aluminum	879
Concrete	850
Diamond	509
Glass	840
Helium	5,193
Water	4,181

Low *vs.* high specific heat

Substances such as metals (e.g., copper) have a *low specific heat* because it does not take much energy to transfer the heat and excite the molecules.

Raising the value of their average kinetic energy (raising their temperature).

Materials that are difficult to heat (e.g., rubber) have a much *higher specific heat* because more energy is required to raise the temperature by the same amount.

Previous table shows that water has a much higher heat capacity than other common materials.

Specific heat modulates environmental temperatures

Raising the ocean by 1 °C takes more energy than raising city sidewalks by the same amount. Thus, land warms faster during the day than water when subjected to the same amount of thermal energy from the Sun. This is the reason for the sea breeze effect.

Hot land heats the air above it, causing it to rise and create a low-pressure area above. The cooler water creates a high-pressure system as air cools and descends. The difference in pressures creates airflow from the sea to the land during the day as a sea breeze.

At night, the land cools faster than the water (i.e., it has a lower specific heat capacity), and the effect is reversed as air flows from the land to the sea.

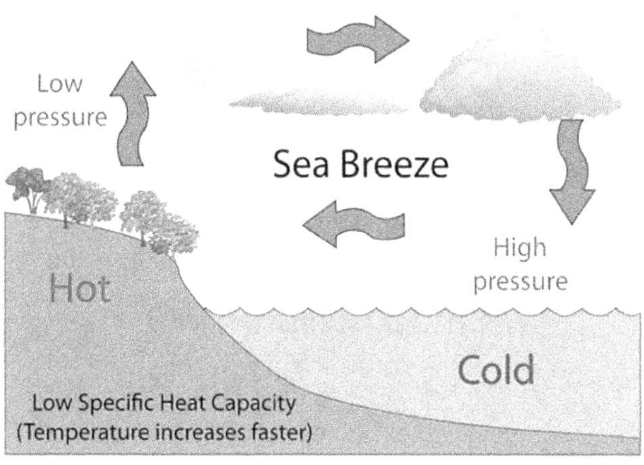

High specific heat of water helps modulate global temperatures

Notes for active learning

Calorimetry

Calories

Calories relate heat to changes in temperature, which is often useful when solving problems in thermochemistry.

calorie (lowercase *c*) is the heat needed to raise the temperature of one gram of water by one degree Celsius (g/°C).

$$1\,cal = 1\,\frac{g}{°C} = 4.19\,J$$

Nutritional Calories

Food Calories (uppercase *C*) are not equivalent to heat calories. This resembles the relationship between kilograms and grams (1 Calorie = 1,000 calories).

Another difference is that while nutritional Calories measure stored energy just as calories do, they also measure the energy stored in the chemical bonds of food that are broken down and stored during digestion. Calories measure the energy in a substance and can be converted to equivalent forms of conventional energy.

For example, the calories contained in 5 pounds of spaghetti have enough energy to brew a pot of coffee; the calories in one piece of cheesecake can power a 60-watt incandescent light bulb for 1.5 hours; 222 Big Macs contain enough energy to drive a vehicle for almost 90 miles.

Some useful conversion relationships:

1 calorie = 4.2 J

1 Calorie (with capital C) = 1,000 calories = 1 kilocalorie

1 Calorie = 4,200 J

Nutrient Molecule in Food	Example	Cal/g	kJ/g
Carbohydrate	Table sugar, potatoes, flour	4	17
Protein	Meats, fish, beans	4	17
Fat	Oil, butter	9	38

Foods contain varying amounts of energy, as shown

Calorimeters

Calorimetry is an experimental method to calculate enthalpy changes by measuring heat flow.

Calorimeters measure the amount of energy in food samples experimentally. Food is placed inside a closed container surrounded by water and burned in the presence of oxygen. Water temperature change is related to the energy released, commonly reported in nutritional Calories (or kilojoules). Note the term *Calories* in uppercase.

Even though cells within the body combust molecules differently from a calorimeter, the calorimeter provides accurate caloric values because it produces the same end products.

Calibration is an experiment in which hot water of known mass and temperature is poured into an insulated cup containing icy water of known mass and temperature. The sample equilibrates to a maximum temperature, and a constant value for the calorimeter is determined below.

Calorimeters are ideally considered an insulated system. The heat liberated by the reaction process is transferred to the other substances within the calorimeter or the calorimeter itself. No heat is lost to the surroundings.

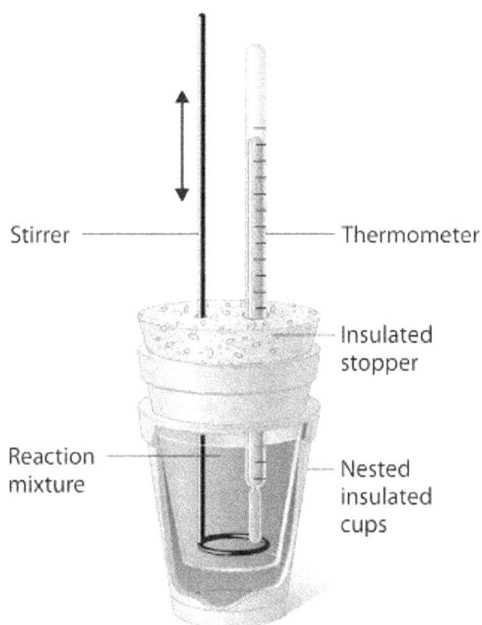

Components of a simple calorimeter

Since no thermal energy is lost in a calorimeter, the heat of a reaction can be measured by inserting a thermometer into a calorimeter and recording the temperature readings before and after the reaction. A stirrer ensures that all calorimeter contents are mixed and react uniformly.

Constant-pressure calorimeters

Constant-pressure and *constant-volume calorimeters* are two types of calorimeters.

Constant-pressure calorimeters measure heat flow at a constant atmospheric pressure.

A perfect calorimeter does not exist; no calorimeter completely insulates heat flow. The calorimeter itself, as part of the surroundings, absorbs some heat. Therefore, the calorimeter must be calibrated to obtain accurate results.

Styrofoam cups are simple, constant-pressure calorimeters often used in introductory chemistry classes. Although a Styrofoam cup's temperature-insulating properties are reasonably good, the cup absorbs heat.

Calorimeter constant K_{cal}

Calorimeter constant K_{cal} is determined by a composite sample equilibrating at maximum temp.

Principle of calorimetry:

heat lost by hot water = heat gained by cool water + heat gained by calorimeter

Determine the mathematical equations:

$$-m_{hot}c_{water}(T_{final} - T_{intial, hot}) = m_{cool}\, c_{water}(T_{final} - T_{initial, cool}) + K_{cal}\,(T_{final} - T_{initial, cool})$$

Sign is negative for heat lost by the hot water because heat loss is a negative value. Substitute known masses and temperatures into the expression. Since the cool water was in the cup, the calorimeter's initial and cool temperatures were the same.

For example, suppose 60.1 g of water at 97.6 °C is poured into a cup calorimeter containing 50.3 g of water at 24.7 °C. The final temperature of the combined water samples reaches 62.8 °C. What is the calorimeter constant? (Use c_{water} = 4.184 J/g°C)

$$-(60.1\text{ g})\cdot(4.184\text{ J/g°C})\cdot(62.8\text{ °C} - 97.6\text{ °C})$$
$$= [(50.3\text{ g})\cdot(4.184\text{ J/°C})\cdot(62.8\text{ °C} - 24.7\text{ °C})] + [K_{cal}\,(62.8\text{ °C} - 24.7\text{ °C})]$$

$$-(60.1\text{ g})\cdot(4.184\text{ J/g°C})\cdot(-34.8\text{ °C})$$
$$= [(50.3\text{ g})\cdot(4.184\text{ J/°C})\cdot(38.1\text{ °C})] + [K_{cal}\,(38.1\text{ °C})]$$

$$[8.75 \times 10^3\text{ J}] - [8.02 \times 10^3\text{ J}] = [K_{cal}\,(38.1\text{ °C})]$$

$$0.73 \times 10^3\text{ J} = [K_{cal}\,(38.1\text{ °C})]$$

$$[0.73 \times 10^3\text{ J}] / (38.1\text{ °C}) = K_{cal}$$

$$K_{cal} = 19.2\text{ J/°C}$$

Heat flow

After calibrating the calorimeter, the specific heat capacity will be determined.

A known quantity of metal is heated to a known temperature and placed in an insulated cup calorimeter containing a known quantity and temperature until the temperature equilibrates.

For example, suppose 28.2 g of an unknown metal is heated to 99.8 °C and placed into a cup calorimeter containing 150.0 g of water at 23.5 °C. The temperature of the water equilibrates at a final temperature of 25.0 °C. What is the specific heat capacity of the metal? (Use the calorimeter constant of 19.2 J/°C).

Consider heat flow:

$$\text{heat lost}_{\text{hot metal}} = \text{heat gained}_{\text{water}} + \text{heat gained}_{\text{calorimeter}}$$

$$-m_{hot}C_{water}(T_{final} - T_{intial,\ hot}) = m_{cool}\ c_{water}(T_{final} - T_{initial,\ cool}) + K_{cal}\ (T_{final} - T_{initial,\ cool})$$

$$-(28.2\text{ g})\cdot(C_{metal})\cdot(25.0\ °C - 99.8\ °C) =$$
$$= [(150.0\text{ g })\cdot(4.184\text{ J/g°C})\cdot(25.0\ °C - 23.5\ °C)] + [19.2\text{ J/°C}(25.0\ °C - 23.5\ °C)]$$
$$-(28.2\text{ g})\cdot(C_{metal})\cdot(-\ 74.8\ °C)$$

$$= [(150.0\text{ g })\cdot(4.184\text{ J/g°C})\cdot(1.5\ °C)] + [19.2\text{ J/°C}(1.5\ °C)]$$

$$2.11 \times 10^3\text{ g°C }(C_{metal}) = (9.4 \times 10^2\text{ J}) + (29\text{ J})$$

$$2.11 \times 10^3\text{ J }(C_{metal}) = 969\text{ J}$$

Solving for specific heat capacity:

$$C_{metal} = 969\text{ J} / 2.11 \times 10^3\text{ g°C}$$

$$C_{metal} = 0.459\text{ J/ g°C}$$

Constant-volume calorimeters

Constant-volume calorimeter is used to measure heat flow at constant volume.

Bomb calorimeter (*decomposition vessel*) is the most used constant-volume calorimeter.

Combustion is when oxygen reacts with a substance to produce carbon dioxide (CO_2), water (H_2O), and energy.

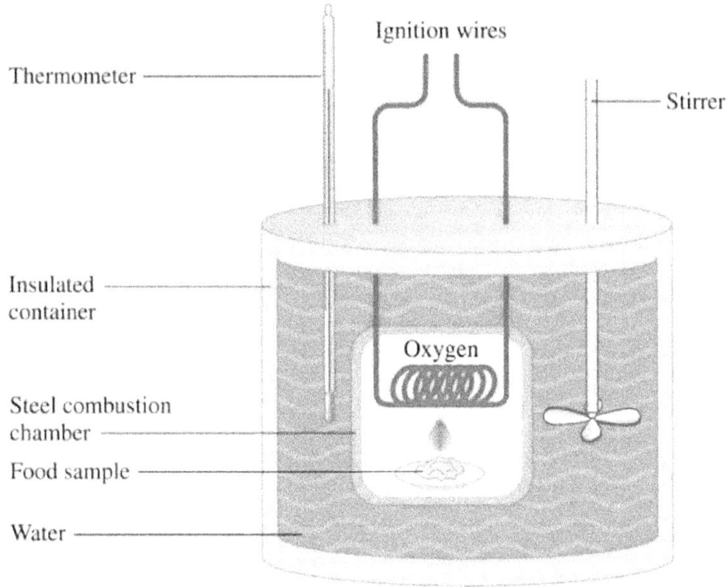

Bomb calorimeter: constant-volume calorimeter

Notes for active learning

Heat Transfer

Rate of heat transfer

When heat is transferred by conduction, the rate of heat transfer (H) can be calculated in joules per second (J/s) or watts (W).

Amount of energy (heat) conducted per unit:

$$H = \frac{kA\Delta T}{t}$$

where ΔT is the temperature difference across the object (K), A is the cross-sectional area (m^2), t is the thickness of the material (m), and k is thermal conductivity of the material (J/s·m· K).

Heat transfer (H) rate is a material characteristic that is usually provided in the problem.

Heat transfers between two objects by *conduction*, *convection*, or *radiation.*

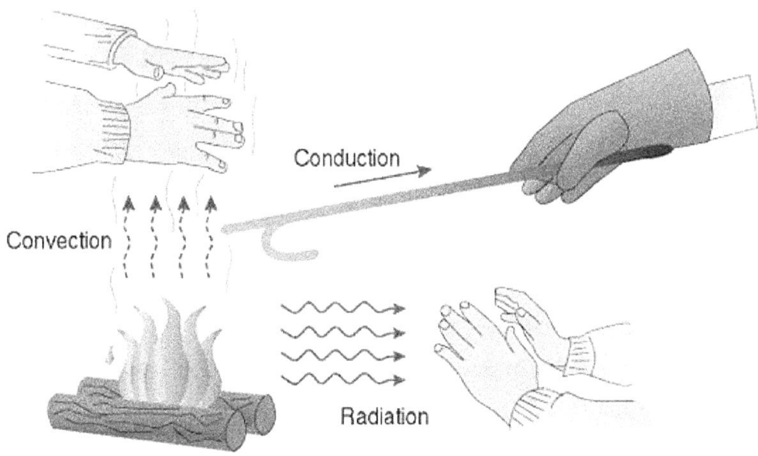

Heat transfer by convection, conduction, and radiation

Heat flows from hotter object

According to the First Law of thermodynamics, *heat always flows from a hotter to a colder system until the two systems are in thermal equilibrium.*

Conduction is the flow of heat between two materials in *direct contact.*

Convection is the flow of energized (e.g., air) molecules that settle and transfer their energy by *direct contact* using conduction.

Radiation is the heat flow (i.e., energy) emitted by matter through *electromagnetic radiation* (e.g., photons or electromagnetic waves).

Thermal conductivities of selected materials

Material	k (J/s·m·K)
Aluminum	237
Concrete	1
Copper	386
Glass	0.9
Stainless Steel	16.5
Water	0.6

Thermal conductivities of common materials

Conduction transfers heat by contact

Conduction is a form of heat transfer that occurs between substances in direct contact (i.e., substances that are touching).

When particles of a hotter substance vibrate, these molecules bump into nearby particles and transfer some energy.

Conduction is an essential means of heat transfer for solids, and metals are conductive. It occurs when heat is transferred through *direct contact.*

For example, on a frigid winter night, when a person wraps their hands around a cup of hot cocoa, their fingers become warm because they are in direct contact with the hot mug, which is in direct contact with the hot liquid. The same process is happening at a molecular level.

Liquid in the mug contains heat energy, and the molecules of the hot liquid are in constant motion (high kinetic energy). These molecules continuously collide with the molecules of the mug, increasing their motion. The collisions gradually transfer energy from the hot liquid molecules to the mug molecules.

The mug transfers heat to a person's hands through the same conduction mechanism. The person perceives energy transfer as increasing temperature; thus, their hands feel warm.

Calculating heat conduction

When calculating heat conduction through an object, it is essential to note that the cross-sectional area refers to where the heat is being conducted. For the mug example, the heat is being conducted through its outer curved surface.

When the mug is placed on a table and the heat conduction to the table is calculated, the cross-sectional area is the area of the mug's bottom, not the area of the sidewalls. Therefore, the cross-sectional area is the circumference of the mug times its height. The thickness, in this case, is the thickness of the mug itself.

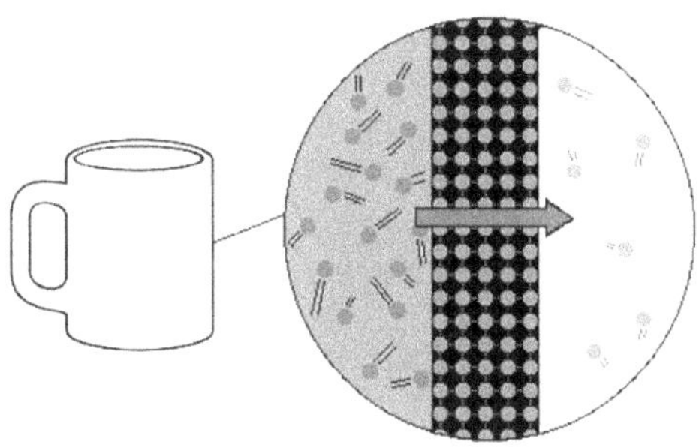

Conduction (heat transfer by direct contact) from the hot liquid to the walls of a mug

Convection transfers heat by motion of particles

Convection refers to the *physical flow of matter*. It occurs when heat flows from one place to another through the movement of fluids, energizing molecules along the way.

Convective heat transfer occurs by *diffusion* (i.e., random Brownian motion of particles) and advection (i.e., larger-scale flowing currents transport matter or heat).

For example, convection occurs when warmer areas of a liquid or gas rise to colder areas, and the cooler liquid or gas replaces the warmer areas.

Gas is fluid, as it can flow and change shape depending on its container.

A pot of water is placed on a hot stove. The water at the bottom of the pot warms first. As the water warms, it rises and is replaced by cooler water. This circulation of the heated water transfers thermal energy from the bottom of the pot to the top.

This movement of molecules is a closed fluid flow pattern of *convection currents*.

Convection ovens work similarly with gas instead of liquid molecules. Air is blown past a heating element to the oven chamber. The hot element energizes the air molecules as they pass through the heating element. Energized air molecules are circulated throughout the oven by airflow provided by the fan and continue until all molecules are energized to the same amount.

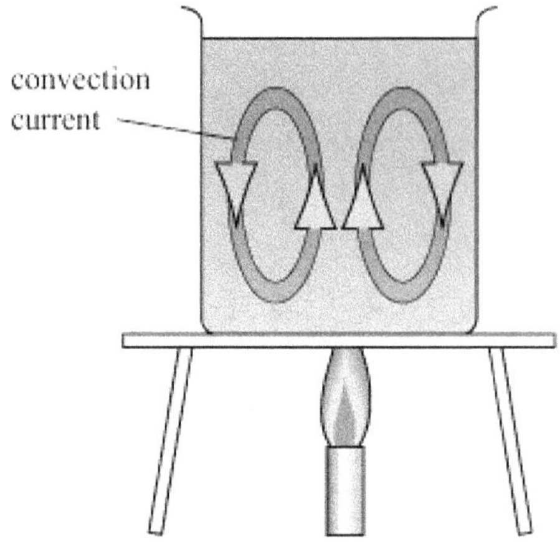

Convention current represented by the circulating water

Radiation transfers heat through electromagnetic waves

Radiation transfers heat *without* contact between the heat source and the heated object. It consists of electromagnetic waves traveling at the speed of light and can occur in a vacuum.

It has *no mass exchange*, and *no medium is required* to transfer heat.

Radiation transfers energy through electromagnetic waves emitted from all objects or surfaces with heat energy and does not require a medium; thus, radiation can occur in a vacuum.

Objects with more energy emit more radiation and do not require a medium; thus, radiation can occur in a vacuum.

For example, a hot piece of metal radiates heat as infrared electromagnetic waves. As the metal gets hot, it also radiates heat in the visible spectrum, thus making the metal appear “red hot.”

Sun is a radiation source, releasing electromagnetic waves through space and the atmosphere to heat Earth. Similarly, holding hands next to a fire heats hands through radiation.

Radiation can be important even in situations with an intervening medium. An example is the heat transfer between a living entity and its surroundings.

For example, the sun's thermal energy travels through the 93 million miles of space to warm the Earth, even passing through the mesosphere, the coldest part of the Earth's atmosphere, which can reach temperatures as low as -173 °C.

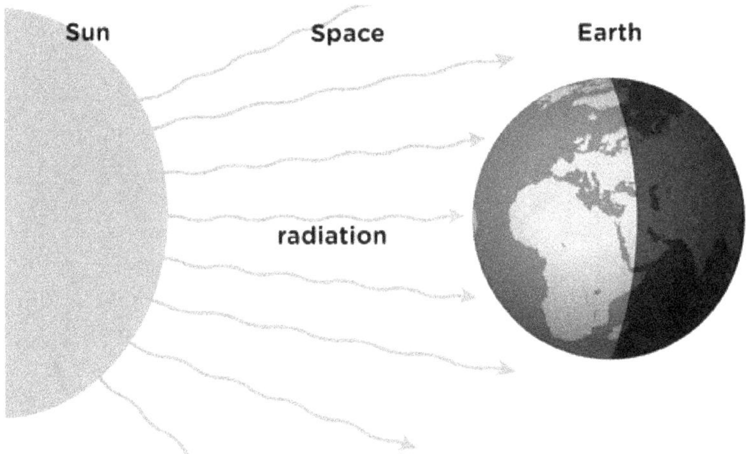

Radiation emitted from the Sun travels to the Earth's surface

Power radiated by an object is related to its temperature:

$$P = e\sigma A(T^4 - T_C^4)$$

where P is the power radiated (W), e is the emissivity of the radiator, σ is Stefan's constant equal to 5.67×10^{-8} W/m^2·K, A is the area of the radiating surface (m^2), T is the temperature of the radiator (K), and T_C is the temperature of the surroundings (K).

Notes for active learning

Thermodynamic Systems

Thermodynamic variables

Thermodynamics defines variables related to the energy of physical and chemical changes.

Thermodynamic variables:

ΔE or ΔU	measures *total internal energy* of a system
q	measures *heat* of a system
w	measures *work* of a system
ΔH	measures *enthalpy change* of a system – whether a process is exothermic or endothermic
ΔG	measures *Gibbs free energy change* of a system – whether a process is spontaneous or nonspontaneous
ΔS	measures *entropy change* of the system or surroundings – whether the randomness of the system or surroundings is increasing or decreasing

Steady-state processes occur without a change in the internal energy.

System and surroundings

Thermodynamic system is a body of matter or radiation confined by partitions, with defined permeabilities separating it from its surroundings.

System defines the parameters of the universe being studied, such as the substances involved in a chemical reaction or a phase change.

Surroundings are everything other than the system being studied.

Isolated, closed, and open systems

Thermodynamic systems describe the processes and matter within a boundary.

*Isolated system*s have no heat transfer, work, or matter with its surroundings.

Closed systems may transfer heat and work to their surroundings but *not* matter between the system and its surroundings.

Open systems may transfer heat, work, or matter between itself and its surroundings.

Note: *heat* and *temperature* are different quantities.

Heat is a form of *thermal energy* that can *flow into or out of a system.*

Temperature measures the *average kinetic energy* (KE) of particles in a system.

Characteristics of thermodynamic systems

System	Exchanges with surrounding	Total amount of energy	Example	Illustration
Open	Energy & Matter	Not remain constant	Solution in an open flask	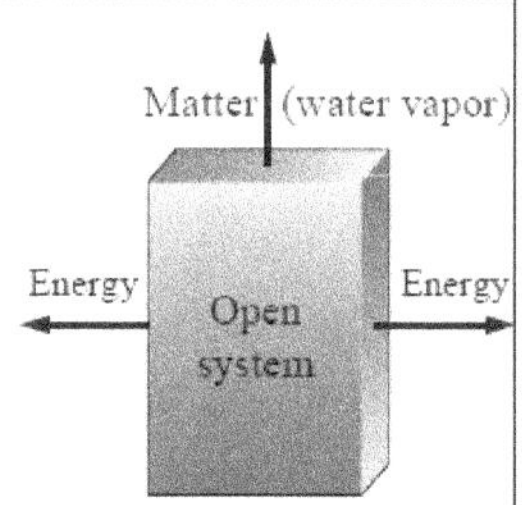
Closed	Only Energy	Not remain constant	Solution in a sealed flask	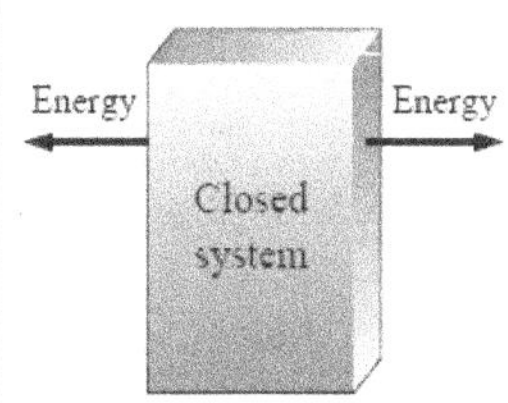
Isolated	Neither energy nor matter	Remains constant	Sealed flask in a thermos flask	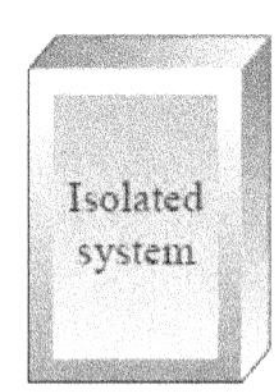

State functions

State function describes the energy required to change a system into another equilibrium state.

State functions (i.e., state quantities) are a group of equations representative of the properties of a system, which depend *only* on the current state of the system, not the path used to achieve the current state.

State functions are quantities with values *that are independent of the path used to measure them.*

State functions ($\Delta E°$, $\Delta H°$, $\Delta S°$, $\Delta G°$) depend only on the *initial and final states* of the system.

State function variables (e.g., pressure, volume, temperature) are functions of *other state variables.*

Thermodynamic state functions include *enthalpy* (H), *entropy* (S), and *Gibbs free energy* (G).

Path functions

Process functions quantitatively describe the transition between equilibrium states of a thermodynamic system.

Path functions depend on the *path taken* to transition from one state to another. Therefore, different routes give different quantities (e.g., work, heat).

Quantities, such as work (w) and heat (q), are not state functions but are *path functions*, represented by *lowercase letters.*

Notes for active learning

Endothermic and Exothermic Reactions

Endothermic *vs.* exothermic processes

Endothermic processes involve the system gaining heat from its surroundings, which is equal to the heat lost by the surroundings.

Exothermic processes cause the system's heat loss to equal the heat gain of the surroundings.

Some chemical reactions produce energy, while others require energy in the form of heat.

Energy is used to *form or break chemical bonds.*

Enthalpy change (ΔH) describes energy changes associated with forming or cleaving a chemical bond.

Unit for enthalpy (H) is the Joule (J).

Joule is the same unit used for energy; it can be expressed as energy per mol (J/mol).

Exothermic reactions release heat

Exothermic reactions are characterized by the fact that the energy needed to initiate a reaction is less than the energy released, resulting in the liberation of *heat.*

Exothermic reactions have a change in *enthalpy of less than zero.*

$$\Delta H < 0$$

Energy is *released* into the surroundings because stronger, more stable bonds form in the products than in the reactants.

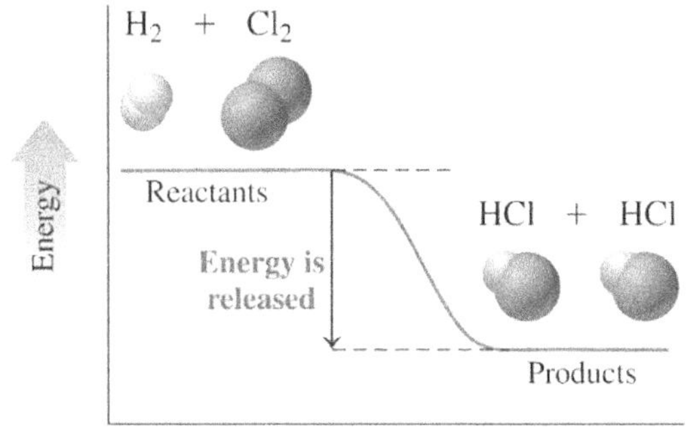

$$H_2\,(g) + Cl_2\,(g) \rightarrow 2\,HCl\,(g) + 185\text{ kJ}$$

In exothermic reactions, heat is released, and the sign of ΔH is negative ($-\Delta H$); the energy of products is *less* than the energy of the reactants, as determined by bond-dissociation energies.

energy is released $\rightarrow \Delta H < 0$

$\Delta H < 0 \rightarrow$ enthalpy decreases

Exothermic reactions can be represented in a chemical reaction with *heat* on the product side:

$$A + B \rightarrow C + heat$$

For example, the chemical reaction for hydrogen and chlorine gas is:

$H_2\,(g) + Cl_2\,(g) \rightarrow 2\ HCl\,(g) + 185$ kJ

$\Delta H = -185$ kJ/mol (heat released)

Heat is released (i.e., exothermic reaction), and ΔH is negative ($-\Delta H$).

Another example of an exothermic reaction is the burning of fuel (i.e., combustion).

Endothermic reactions absorb heat

Endothermic reactions have a change in *enthalpy greater than zero.*

$\Delta H > 0$

Endothermic reactions require (net) energy to break bonds.

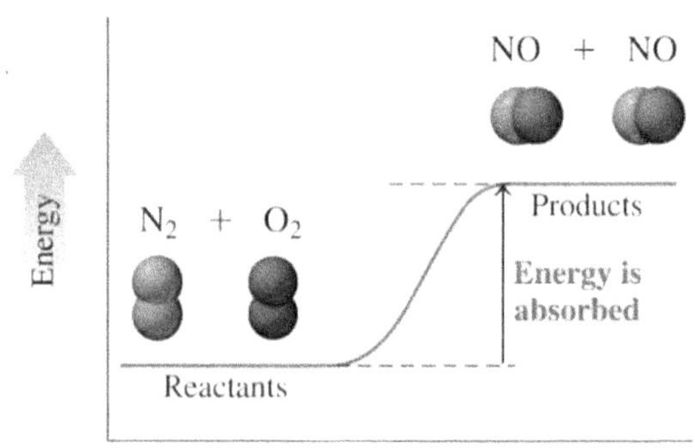

$$N_2\,(g) + O_2\,(g) + 181\ \text{kJ} \rightarrow 2\ NO\,(g)$$

In endothermic reactions, weaker bonds form in the products than in the reactants.

For example, an endothermic chemical reaction represented with heat as a reactant:

$heat + NH_4NO_3\,(s) \rightarrow NH_4^+\,(aq) + NO_3^-\,(aq)$

Above reaction is the ionic compound ammonium nitrate dissolving in water (H_2O), dissociating into constituent ions, NH_4^+ and NO_3^-.

Endothermic reactions include *melting ice* and cooking an egg (*heat is required* for processes).

In an endothermic reaction, heat is absorbed, and the sign of ΔH is ($+\Delta H$); the energy of the products is greater than the energy of the reactants.

Amount of energy absorbed is calculated using bond-dissociation energies.

energy is absorbed $\rightarrow \Delta H > 0$

$\Delta H > 0 \rightarrow$ enthalpy increases

Endothermic reactions can be represented in a chemical reaction with *heat* on the reactant side:

$$A + B + heat \rightarrow C$$

For example, nitrogen gas (N_2) and oxygen gas (O_2) form nitrogen monoxide (NO):

$$N_2\,(g) + O_2\,(g) + 181\text{ kJ} \rightarrow 2\text{ NO}\,(g)$$

$\Delta H = +181$ kJ (heat added)

Heat is absorbed (i.e., endothermic reaction), and ΔH is positive.

Exothermic and endothermic reactions comparison

The figure below depicts an exothermic (producing heat) and endothermic reaction (absorbing).

Reaction is exothermic on the left because heat is *produced and released* into the surroundings.

Reaction is *endothermic* on the right because heat is *absorbed* from the surroundings.

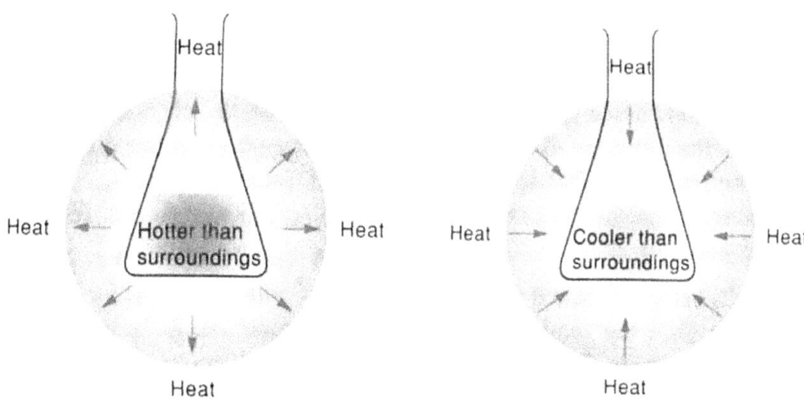

Exothermic process releases heat (left), and the endothermic process absorbs heat (right)

Reaction	**Energy Change**	**Heat in the Equation**	**Sign of ΔH**
Endothermic	Heat absorbed	Reactant side	Positive (+)
Exothermic	Heat released	Product side	Negative (–)

For example, is the following reaction endothermic or exothermic?

$$O_2N–NO_2 \rightarrow O_2N + NO_2$$

Reaction is endothermic (absorbs heat) because it forms two products from a single reactant.

Breaking chemical bonds *require energy*.

Thermodynamics

Heat, work and temperature

Thermodynamics relates heat, work, and temperature to energy, entropy, and properties of matter and radiation.

Thermodynamics answers several fundamental questions regarding *whether* a reaction occurs, to what *extent*, and whether it *releases or absorbs heat.*

An important concept in thermodynamics is the thermodynamic system, which refers to the region under study. Everything else is the surroundings.

Systems are separated from the surroundings by a boundary. Transfers of energy as work, heat, or matter between the system and surroundings occur through the boundary.

Four laws of thermodynamics

Thermodynamics states four laws that characterize the relationship between heat and energy, describing behavior of physical quantities such as temperature, heat, work, energy, and entropy.

> *Zeroth Law of Thermodynamics* (or *concept of temperature*) states that ***heat flows from the hotter body to the colder*** *one when they are placed in contact, until thermal equilibrium is reached.*

> *First Law of Thermodynamics (or the Law of Conservation of Energy) states that the total energy in a system remains constant, as* ***energy can neither be created nor destroyed****.*

> *Second Law of Thermodynamics* (or *concept of entropy*) states that *the universe's natural order* ***favors a net increase*** *in the overall entropy (i.e., disorder).*

> *Third Law of Thermodynamics* (or *entropy at absolute zero*) states that *the entropy of a perfectly ordered, pure crystalline solid is* ***zero at the temperature of absolute zero*** *(equal to 0 K, –273.15 °C or –459.67 °F).*

Notes for active learning

Zeroth Law of Thermodynamics

Thermal equilibrium

Zeroth Law of Thermodynamics is the logical precursor to the First and Second Laws of Thermodynamics.

Zeroth Law states that *if two systems are in thermal equilibrium with a third system, the two initial systems are also in thermal equilibrium.*

Three systems in thermal equilibrium contain the same amount of heat energy (the same temperature) and do not exchange heat, which is the fundamental principle underlying thermodynamics.

For example, if system A is in equilibrium with B, and system C is in equilibrium with B, then systems A and C must be in equilibrium, and all three systems must be at the same temperature.

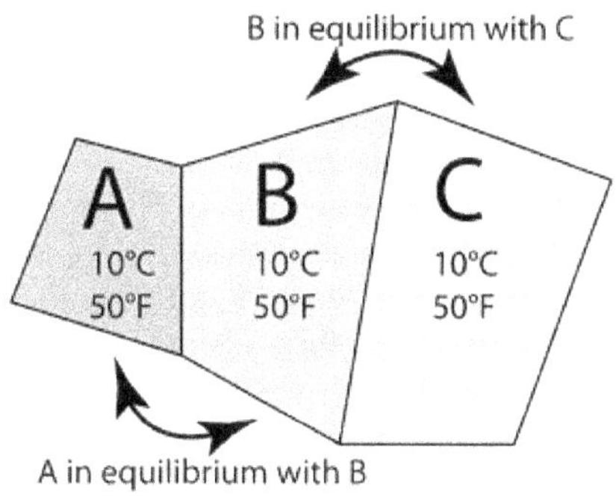

Zeroth Law of Thermodynamics with three systems in thermal equilibrium

However, it is vital to notice that although no net heat transfer occurs between systems in equilibrium, thermal energy is technically transferred between the related systems.

Exchange of thermal energy requires that every unit of energy passed from any system has the same value as the energy passed back into the system. This holds even if the two systems contain materials with different specific heats.

This observation of thermal energy transfer implies that there must be a property that remains constant upon which heat transfer depends, and that property is temperature.

Concept of temperature

Zeroth Law of Thermodynamics states that heat flows from the hotter body to the colder body when they are in contact until thermal equilibrium is reached.

It establishes that *temperature is a core, measurable property of matter.*

Separated by a permeable barrier, two systems are in thermal equilibrium, and *no heat* transfers.

After thermal equilibrium is achieved, no heat transfer occurs, defining *concept of temperature.*

Expressing thermal equilibrium

If two systems are in thermal equilibrium with a third system, then each is in thermal equilibrium with the other.

Thermal equilibrium is expressed by:

$$T_A = T_B$$

and

$$T_C = T_B$$

then

$$T_A = T_C$$

Zeroth Law is crucial for calculations of thermodynamics (i.e., heat and energy relationships).

First Law of Thermodynamics

Conservation of energy

Energy, like matter, is always conserved; it can neither be created nor destroyed.

First Law of Thermodynamics states that *energy can neither be created nor destroyed, and the total energy in a system remains constant. Energy is conserved in a chemical reaction.*

First Law of Thermodynamics relates to the *internal energy* of a substance.

Internal energy increases if heat is added to the system or if work is done on the system.

Disordered motion of particles

Change in total *internal energy* of a system equals contributions by heat (q) and work (w):

$$\Delta U = q - w$$

where ΔU is the change in internal energy, q is the contribution from heat, and w from work.

Internal energy refers to the microscopic energy of the disordered motion of particles and is represented by the symbol (U).

Internal energy cannot be seen and has no apparent effect on the object's motion, but it does play a part in the thermodynamic properties of the substance. This principle is the foundation for the law of *conservation of energy*.

For example, a glass of water does not appear to contain kinetic or potential energy. Water molecules are whizzing around on the atomic scale, moving at hundreds of meters per second. If the water is dumped from the glass, its internal energy does not affect its motion.

Thermodynamic properties of the substance include potential and kinetic energy

Sign conventions

$\Delta U = q - w$ uses the following sign conventions:

q is *positive* if heat is *absorbed into* the system (i.e., heating)

q is *negative* if heat is *released from* the system (i.e., cooling)

w is *positive* if work is done *by* the system (i.e., compression)

w is *negative* if work is done *on* the system (i.e., expansion)

First Law of thermodynamics:

$$\Delta U = q + w$$

where q is *positive* if heat is *absorbed into* the system and *negative* if heat is *released from* the system; work (w) is positive if done *on the* system or negative if done *by* the system.

Heat (q) and work (w) absorbed or released

Use the sign conventions in the chart below:

q	+	Heat ***absorbed*** by the system	E_{system} increases
q	–	Heat ***released*** by the system	E_{system} decreases
w	+	Work done ***on*** the system	E_{system} increases
w	–	Work done ***by*** the system	E_{system} decreases

Heat or energy *flowing out* of the system is *negative.*

Heat or energy *flowing into* the system is *positive.*

Other natural sciences use the opposite sign convention; note the sign convention used.

Internal energy

The difference between the amount of heat added to the system (Q) and work done by the system (W) equals the total change in internal energy (ΔU):

$$\Delta U = \Delta Q - \Delta W$$

As with other forms of energy, the unit of internal energy is the joule (J), although it is also expressed in calories (cal).

By convention, work done *on a system* is positive, and work done *by a system* is negative.

Occasionally, the above equation is written as $\Delta U = \Delta Q + \Delta W$, and ΔW is defined as the work done on the system. Be sure to note which form is being used.

Calculating internal energy

The quantities in $\Delta U = \Delta Q + \Delta W$ can determine what a thermodynamic system is experiencing.

For example, in a chemical reaction, if ΔQ is a quantity less than zero ($\Delta Q < 0$), the reaction is losing heat. This reaction is *exothermic,* meaning that heat is a byproduct.

If ΔQ is a quantity greater than zero ($\Delta Q > 0$), the reaction absorbs heat from its surroundings. This is an *endothermic* reaction, where heat acts as a *reactant.*

Internal energy is expressed:

$$\Delta U = q + w$$

where ΔU is the change in internal energy, q is the contribution from heat, and w from work.

For example, consider internal energy, heat, and work. For a process in which 3.4 kJ of heat flows out of the system while the system in the surroundings does 4.8 kJ of work, what is the internal energy?

Calculate internal energy:

$$\Delta U = q + w$$

$$\Delta U = -3.4 \text{ kJ} + (-4.8 \text{ kJ})$$

$$\Delta U = -8.2 \text{ kJ}$$

where ΔU is the change in internal energy, q is *negative* as heat is *released from* the system, and work (w) is negative when done *by* the system.

Use the appropriate heat (q) and work (w) signs.

Energy has flowed from the system to the surroundings by –8.2 kJ.

Notes for active learning

Second Law of Thermodynamics

Concept of entropy (*S*)

Second Law of Thermodynamics is that *the universe favors a net increase in the overall entropy.*

Isolated systems increase in entropy and *never decrease in entropy.*

Open systems decrease entropy as entropy increases for the surroundings.

Second Law of Thermodynamics states that in an isolated system (where there is no transfer of heat, work, or matter), a process either increases entropy or remains constant, but entropy never decreases.

Entropy is a measure of the randomness or disorder of a system. Entropy measures the system's thermal energy per unit temperature, which is unavailable for useful work.

Work requires ordered molecular motion.

Open systems transfer energy

An exception to the Second Law of Thermodynamics (i.e., entropy cannot decrease) occurs in an open system, where heat, work, or matter can be transferred to the surroundings; the system's entropy decrease is related to an increase in the entropy of the surroundings.

Second Law of Thermodynamics states that *heat flows spontaneously* from a hot object to a cold object, but never spontaneously flows in the opposite direction. A bowl of ice never gets colder in a warm room; the ice always gets warmer.

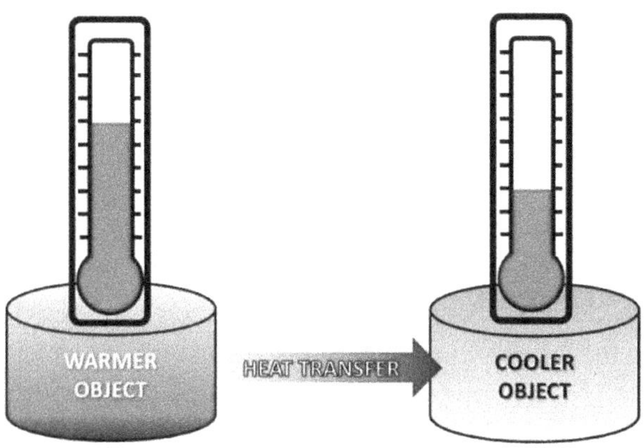

Heat flows spontaneously from a hot object to a cold object

Entropy measures disorder

Entropy measures disorder within a system and likelihood—how likely a system is to remain in the same state twice or how likely a reaction is to occur.

Entropy of a natural system never decreases, as it is the natural tendency of all things to move towards maximum disorder—maximum entropy.

For example, gas expands to fill space with more degrees of freedom to arrange particles—more *multiplicities*.

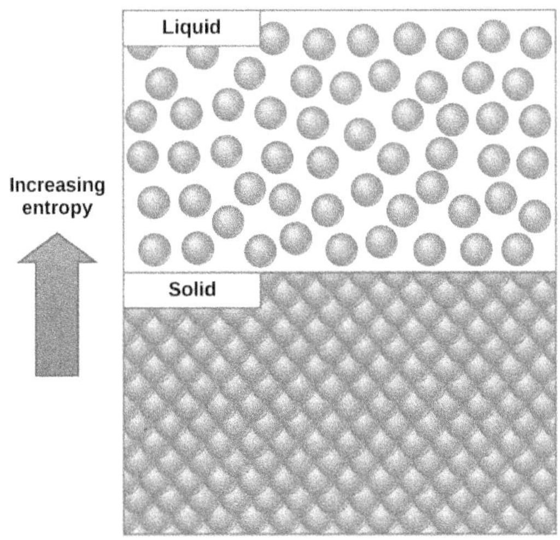

A solid has less entropy than a liquid, which has much less entropy than a gas

According to Second Law of Thermodynamics (i.e., entropy can stay constant or increase but never decrease), entropy is given by:

$$\Delta S \geq \frac{Q}{T}$$

where ΔS is change in entropy of a process (J/K), Q is heat transferred (J), and T is temperature (K).

If a system strives to increase entropy (i.e., disorder), it does not mean that the system cannot become more orderly. This means that to do so, outside energy must be transferred into the system, and it cannot be done on its own.

Nonspontaneous and spontaneous processes

Nonspontaneous processes require energy (or work) for the system to decrease entropy.

Spontaneous processes tend towards randomness without the input of energy, thereby increasing entropy.

For example, a valve connects a container with air and another container in a vacuum. When the valve is opened, air flows from the container to the vacuum in the container. No work or energy is needed for this to happen. The process is *spontaneous*, and entropy *increases*.

For the reverse to occur, energy must be *added to the system* to re-pressurize the original container with air. This process is an example of a *nonspontaneous* reaction.

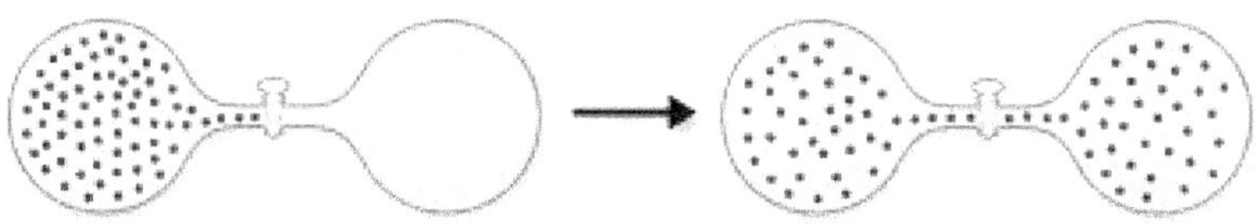

Air flows from the container to the container in a vacuum

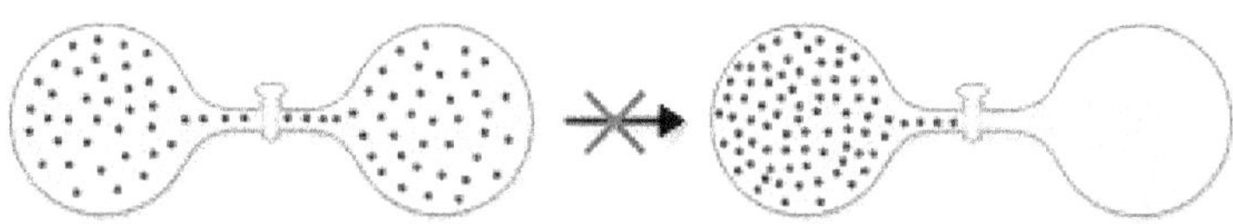

Nonspontaneous reactions do not occur without the input of energy

The universe's entropy always *increases* when a *spontaneous process* occurs.

Entropy of the universe is expressed as:

$$\Delta S_{universe} > 0$$

Reversible and irreversible processes

Thermodynamic processes are *reversible* or *irreversible.*

Reversible processes are idealized, whereas irreversible processes are more common in nature.

Reversible processes

$$\Delta S = q / T$$

Irreversible processes

$$\Delta S > q / T$$

where ΔS is the change in entropy, q is the contribution from heat, and T is the temperature (K).

Natural processes are *irreversible*, so the entropy change is greater than the heat transfer over a given temperature. Due to irreversible nature of processes, the universe's entropy is increasing.

Reactions at equilibrium are *reversible, with no net change to the system or its surroundings* and proceed in both directions.

Reversible processes reverse direction if some change is made to some system property:

$$\Delta S_{total} = \Delta S_{system} + \Delta S_{surroundings}$$

ΔS of the system equals in magnitude, but *opposite in sign*, to the ΔS of the surroundings.

Entropy related to heat and enthalpy

Entropy change of surroundings is related to heat (q) flow, enthalpy (ΔH), and temperature:

$$\Delta S_{surroundings} = -\,q\,/\,T$$

or (at constant pressure)

$$\Delta S_{surroundings} = -\,\Delta H\,/\,T$$

Combining these two equations:

$$\Delta S_{surroundings} = \Delta S_{system} - (\Delta H_{system}\,/\,T)$$

where ΔS is the change in entropy, ΔH *is* the enthalpy, q is the contribution from heat, and T is the temperature (K).

ΔS from tabulated values

ΔS is calculated from standard molar entropies, as for ΔH and ΔG, by heats of formation.

Equation to calculate ΔS from tabulated values:

$$\Delta S^\circ_{rxn} = \Sigma\, n(\Delta S^\circ_f \text{ of products}) - \Sigma\, n(\Delta S^\circ_f \text{ of reactants})$$

where n represents moles of each reactant (or product) as a balanced chemical equation, and ΔS is the change in entropy.

For example, determine the entropy change (ΔS) for the above chemical equation as (ΔH):

$$16\ H_2S\ (g) + 8\ SO_2\ (g) \rightarrow 16\ H_2O\ (l) + 3\ S_8\ (s)$$

ΔS°_{rxn} as a measure of the entropy change of a system:

$$\Delta S^\circ_{rxn} = [(16\text{ mol})\cdot(S^\circ_f \text{ of } H_2O\ (l)) + (3\text{ mol})\cdot(S^\circ_f \text{ of } S_8\ (s))] - [(16\text{ mol})\cdot(S^\circ_f \text{ of } H_2S\ (g)) + (8\text{ mol})\cdot(S^\circ_f \text{ of } SO_2\ (g))]$$

From the table of standard *entropies of formation*, substitute:

$$\Delta S^\circ_{rxn} = \Sigma\, n(\Delta S^\circ_f \text{ of products}) - \Sigma\, n(\Delta S^\circ_f \text{ of reactants})$$

$$\Delta S^\circ_{rxn} = [(16 \text{ mol})\cdot(69.91 \text{ J/mol}\cdot\text{K}) + (3 \text{ mol})\cdot(31.80 \text{ J/mol}\cdot\text{K})] - [(16 \text{ mol})\cdot(205.79 \text{ J/mol}\cdot\text{K}) + (8 \text{ mol})\cdot(161.92 \text{ J/mol}\cdot\text{K})]$$

$$\Delta S^\circ_{rxn} = [1{,}118 \text{ J/K} + 95.40 \text{ J/K}] - [3{,}292.6 \text{ J/K} - 1{,}295.4 \text{ J/K}]$$

$$\Delta S^\circ_{rxn} = -3{,}375 \text{ J/K}$$

where ΔS°_{rxn} is a measure of the entropy change of a system, and the negative sign accounts for the heat of the surroundings, opposite in sign to the heat of the system.

Entropy calculations

For example, the reaction enthalpy (ΔS_{surr}) was –1,876 kJ at a standard temperature of 298 K. The entropy change was calculated as –3,375 J/K (or –3.375 kJ/K). What is the entropy change of the surroundings?

Using the equation:

$$\Delta S_{surroundings} = -\,\Delta H \,/\, T$$

$$\Delta S_{surroundings} = -(-\,1{,}876 \text{ kJ}) \,/\, 298 \text{ K}$$

$$\Delta S_{surroundings} = 6.30 \text{ kJ/K or } \Delta S_{surroundings} = 6.30 \times 10^3 \text{ J/K}$$

where $\Delta S_{surroundings}$ is change in entropy, ΔH is change in enthalpy, and T is temperature (K).

As a *positive* value, the entropy of the surroundings *increased.*

Total entropy change is the sum of the entropy change of the system and the surroundings:

$$\Delta S_{total} = \Delta S_{system} + \Delta S_{surroundings}$$

$$\Delta S_{total} = -3.375 \times 10^3 \text{ J/K} + 6.30 \times 10^3 \text{ J/K}$$

$$\Delta S_{total} = 2.92 \times 10^3 \text{ J/K}$$

where ΔS_{total} is the change in total entropy, ΔS_{system} is the entropy change of the system, and $\Delta S_{surroundings}$ is the entropy change of the surroundings.

Entropy measures degrees of freedom

Entropy (S) measures *randomness* (or *disorder*) from the number of ways (i.e., *degrees of freedom*) a thermodynamic system can be arranged.

Entropy (or disorder) of the universe is *increasing.*

Entropy is defined as energy divided by temperature; the unit is Joules per Kelvin (J/K).

Entropy is expressed as *entropy per unit mass* (J/K·kg) or *unit amount of substance* (J/K·mol).

For example, imagine taking a piece of paper, tearing it into small pieces, and throwing them in the air. How much energy is needed to restore it to its original form?

Consider tearing paper as a *spontaneous* process and putting it together as a *nonspontaneous* process.

Temperature and entropy relationship

The higher the entropy and lower the energy (i.e., temperature), the more stable the system.

Higher entropy + lower energy (lower temperature) → *more stability*

Lower entropy + higher energy (higher temperature) → *less stability*

As *temperature increases*, *entropy increases*:

$T = 0$ K

particles are in equilibrium lattice positions, and S is relatively low

$T > 0$ K

molecules vibrate, S increases

As T increases, more violent vibrations occur, and the entropy (S) increases.

Entropy for gases, liquids, and crystals

Entropy differences between phases of matter (i.e., solid, liquid, gas) are related to the *degrees of freedom* of kinetic energy (KE).

Solids have low entropy because particles forming a crystal solid are packed tightly.

Particles of a solid are confined to vibrating in fixed positions and limited in *degrees of freedom*.

Liquids are intermediate in entropy as intermolecular forces hold particles as a liquid but have increased *degrees of freedom* to move.

Gases are high in entropy because the gas particles have complete *freedom* to move, with a few particles arranged in an orderly fashion.

Low-pressure gas experiences little to no intermolecular forces of attraction or repulsion.

Entropy of phase changes

Entropy changes accompanying phase changes can be estimated, at least qualitatively.

> *Freezing* decreases entropy as the relatively disordered liquid becomes a well-ordered solid.
>
> *Boiling* results in a substantial increase in entropy as the liquid becomes a more highly disordered gas.
>
> *Sublimation* (i.e., from a solid to a gas) is the phase transition with the greatest entropy change for a substance.

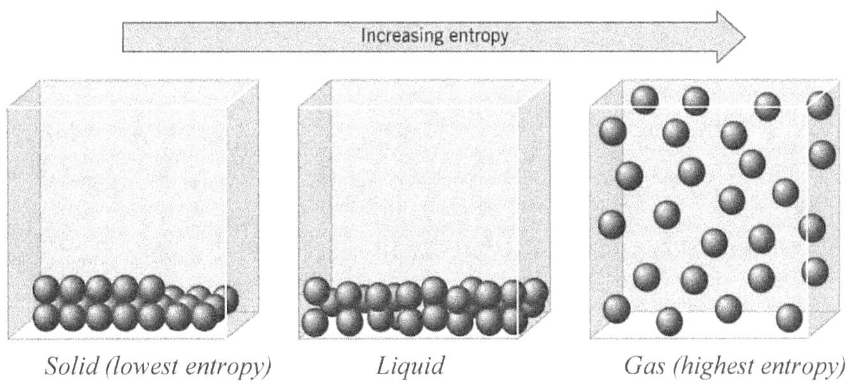

The figure displays a crystalline solid, a liquid, and a gas, with entropy (i.e., randomness) increasing from left to right.

Entropy for gases

Entropy *increases* as gas volume *increases* (i.e., decreasing pressure).

Consider a gas separated from a vacuum by an impermeable partition (figure on the left). Removing the partition (center image) enables more efficient energy distribution.

Gas expands to achieve a higher particle distribution (figure on the right).

The configuration of gas molecules is random and exhibits more positive S (greater disorder).

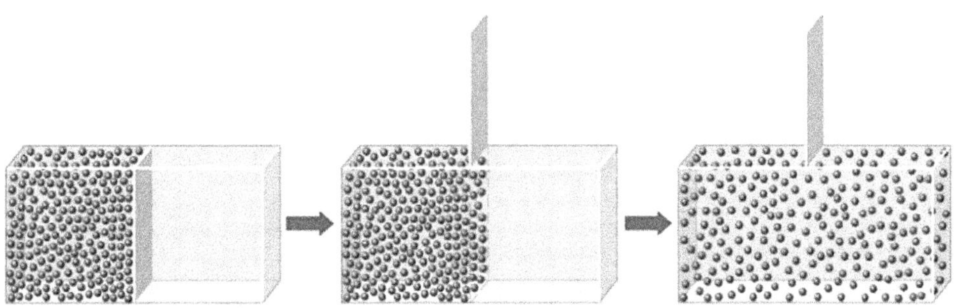

Gas particles are highly ordered on the left (low S) and random on the right (high S)

Number of particles affects entropy reactions, producing more particles as a *positive* ΔS.

Adding particles to a system (as shown on the right below) increases the number of ways energy can be distributed.

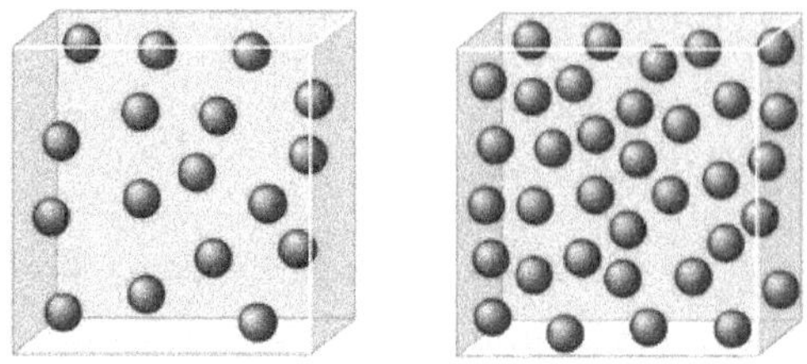

Low entropy *High entropy: more disorder*

Third Law of Thermodynamics

Absolute zero

Third Law of Thermodynamics states that as the temperature of a system approaches absolute zero, its *entropy approaches a constant value.*

Typically, this constant value is zero. In this definition, the entropy of a system is related to the number of microstates possible.

Microstates refer to the atomic configurations possible with the equation:

$$S = k \cdot ln\,(\Omega)$$

where S is the entropy (J/K), k is Boltzmann's constant (1.381×10^{-23} J/K), and Ω is the number of microstates.

Only one microstate is possible at absolute zero, and so Ω equals 1:

$$S = k \cdot ln(1) = k \cdot 0 = 0$$

The entropy of a system at absolute zero is zero.

Minimum energy state

However, some systems have more than one minimum energy state as the temperature approaches absolute zero: the entropy levels off at a value other than zero.

Cooling any process to absolute zero in a finite number of steps is impossible.

In the left image, T = 0 can be reached by taking an infinite number of steps (indicated by the step lines between X_1 and X_2).

In the right image, *a finite number of steps reach $t = 0$*, each approaching zero but never reaching it.

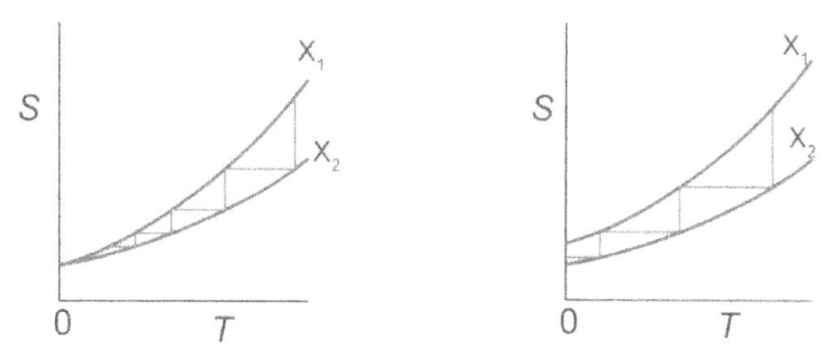

Entropy vs. temperature, illustrating the steps to approach absolute zero

Entropy at absolute zero

Third Law of Thermodynamics states that *the entropy of a perfectly ordered, pure crystalline solid is zero at absolute zero.*

Absolute zero (the coldest temperature possible on the Kelvin scale, which equals 0 K, –273.15 °C, or –459.67 °F), the particles' average kinetic energy (KE) equals zero.

Particles' average kinetic energy (KE) increases as entropy increases.

With increasing temperature (i.e., kinetic energy, KE), substances transition from a solid to a liquid and then to a gaseous state.

For a substance, the gas state has the highest entropy.

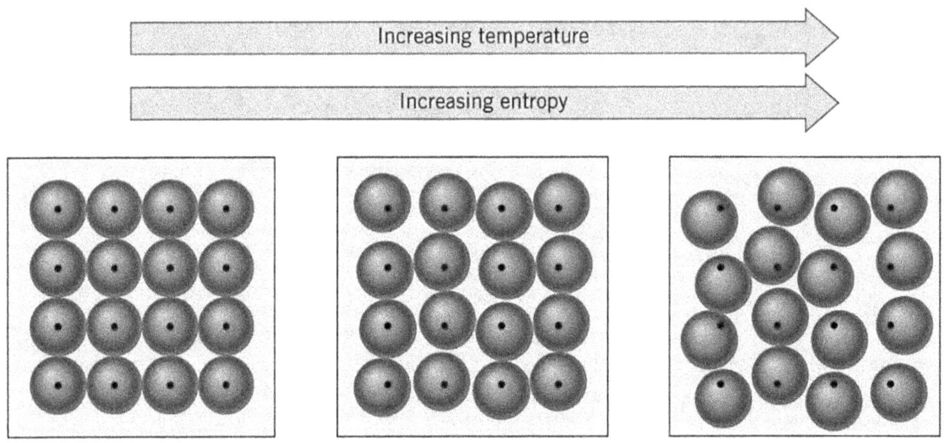

Crystalline solids have low KE with particles arranged orderly.

Liquid particles vibrate from higher KE, and entropy (S) is moderate.

Gas particles have large KE, and entropy (S) is large.

Determining ΔS for reactions

For example, determine the sign of ΔS for the reaction:

$$CaCO_3\,(s) + 2\,H^+\,(aq) \rightarrow Ca^{2+}\,(aq) + H_2O + CO_2\,(g)$$

$$\Delta n_{gas} = 1\text{ mol} - 0\text{ mol}$$

$$\Delta n_{gas} = 1\text{ mol}$$

Since Δn_{gas} is positive, ΔS is positive.

For example, determine the sign of ΔS for the reaction:

$$2\,N_2O_5\,(g) \longrightarrow 4\,NO_2\,(g) + O_2\,(g)$$

$$\Delta n_{gas} = 4\text{ mol} + 1\text{ mol} - 2\text{ mol}$$

$$\Delta n_{gas} = 3\text{ mol}$$

Since Δn_{gas} is positive, ΔS is positive.

For example, determine the sign of ΔS for the reaction:

$$OH^-\,(aq) + H^+\,(aq) \longrightarrow H_2O$$

$$\Delta n_{gas} = 0\text{ mol}$$

$$\Delta n_{gas} = 1\text{ mol} - 2\text{ mol}$$

$$\Delta n_{gas} = -1\text{ mol}$$

Since Δn is negative, ΔS is negative.

Notes for active learning

Enthalpy

Energy changes

Enthalpy (ΔH) of a system is the amount of heat added at constant pressure in a system or reaction and is given in SI units of Joules.

Energy changes in the system can be written as:

$$\Delta U = \Delta H - P\Delta V$$

where ΔH is change in system enthalpy (J), P is pressure (Pa), and ΔV is volume change (m^3).

During an isobaric process (pressure is held constant), such as that conducted in an open container, the change in enthalpy equals the amount of heat transferred during the process. Most reactions do not generate many gaseous products, so little work is associated with the reaction.

Energy change of the system equals the change in enthalpy:

$$\Delta E \approx \Delta H$$

Enthalpy is equal to the transfer of thermal energy (heat); it is only valid at *constant pressure.*

Enthalpy changes can occur at standard temperatures and pressures.

Standard state refers to substances in their most stable form (i.e., the lowest-energy state).

Standard state in thermodynamics refers to a pressure of one atmosphere (1 atm) and a temperature of 25 °C (298 K).

For example, oxygen is O_2 (diatomic gas) at the standard state, and carbon is C (solid graphite).

Standard heats of reaction (ΔH)

Standard heat of reaction (*standard enthalpy of reaction* $\Delta H°_{rxn}$) is the change in heat content when one mole of matter is transformed by a chemical reaction under standard conditions (1 atm and 298 K).

Theoretical and experimental methods calculate the *enthalpy change* of a chemical reaction.

General equation to calculate the heat of a reaction (ΔH) from the difference (delta or Δ) in the energy of the reactants and products is:

$$\Delta H = H_{\text{products}} - H_{\text{reactants}}$$

Standard heats of formation ($\Delta H°_f$)

Standard heat of formation (*standard enthalpy of formation* $\Delta H°_f$) is the change in heat content when one mole of a compound or molecule forms from constituent elements in standard states.

Standard enthalpy of formation has been determined for many molecules, and the $\Delta H°_f$ for the most common molecules are reported and tabulated data.

When choosing $\Delta H°_f$ from a table, choose the value corresponding to the appropriate state of matter (solid, liquid, aqueous, gas).

Standard heat of formation for an element in its standard state equals 0 kJ/mol.

For example, calculate the heat of reaction from the heat of formation and determine if the process is exothermic or endothermic for the balanced chemical equation:

$$16\ H_2S\ (g) + 8\ SO_2\ (g) \rightarrow 16\ H_2O\ (l) + 3\ S_8\ (s)$$

From thermodynamic quantities, use values for the heat of formation of each component in the reaction; set the equation to calculate the heat of the reaction:

$$\Delta H = H_{products} - H_{reactants}$$

$$\Delta H°_{rxn} = [16\ mol(\Delta H°_f\ of\ H_2O\ (l)) + 3\ mol(\Delta H°_f\ of\ S_8\ (s))] - [16\ mol(\Delta H°_f\ of\ H_2S\ (g)) + 8\ mol(\Delta H°_f\ of\ SO_2\ (g))]$$

$$\Delta H°_{rxn} = [16\ mol(-285.8\ kJ/mol) + 3\ mol(0\ kJ/mol)] - [16\ mol(-20.2\ kJ/mol) + 8\ mol(-296.8\ kJ/mol)]$$

$$\Delta H°_{rxn} = [-4{,}573\ kJ + 0\ kJ] - [-323\ kJ + (-2{,}374\ kJ)]$$

$$\Delta H°_{rxn} = -4{,}573\ kJ + 323\ kJ + 2374\ kJ$$

$$\Delta H°_{rxn} = -1{,}876\ kJ$$

$\Delta H°_{rxn}$ is negative, indicating that the reaction is exothermic; the system releases heat to its surroundings.

Enthalpy for exothermic and endothermic reactions

Exothermic reactions have a change in enthalpy of less than zero.

$$\Delta H < 0$$

Endothermic reactions have a change in enthalpy greater than zero.

$$\Delta H > 0$$

Since enthalpy (H) is a state function, it represents a system property; it can be used to describe a reaction or an equation for a reaction.

Calculating the heat of reaction (ΔH)

Calculate the heat of reaction (ΔH) using these steps:

1) Identify the given and needed quantities.
2) Write an expression using the heat of reaction and molar mass needed.
3) Write the conversion factors, including heat of the reaction.
4) Set the expression to calculate heat.

Hess' Law for heat of summation

In 1840, Swiss-born Russian chemist Germain Hess (1802-1850) published Hess' Law of constant heat summation.

Hess' Law describes an essential relationship in physical chemistry that *total enthalpy change (ΔH) during the complete course of a chemical reaction is independent of the number of steps.*

Hess' Law states that the change of enthalpy in a reaction is equal to the sum of the products' enthalpy minus the sum of the reactants' enthalpy:

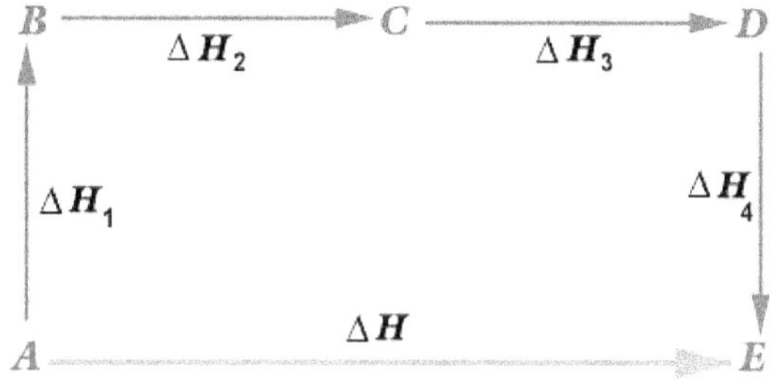

$$\Delta H_{rxn} = \Sigma H_{products} - \Sigma H_{reactants}$$

Enthalpy is a *state function* (i.e., depends only on initial and final states, *not* the reaction's path).

Hess' Law of Heat Summation is expressed

$$\Delta H^\circ_{rxn} = \Sigma\, n(\Delta H^\circ_f \text{ of products}) - \Sigma\, n(\Delta H^\circ_f \text{ of reactants})$$

where n represents the moles of each reactant or product in the balanced chemical equation, and the Greek letter Σ means the sum of the variables.

Hess' Law states that the total energy change throughout a reaction is the same, regardless of whether the reaction occurs in one step or several steps.

Calculating heat of summation

For example, how much heat (kJ) is absorbed when 1.65 grams of NO gas are produced in the following reaction?

$N_2\,(g) + O_2\,(g) \rightarrow 2\ NO\,(g)$

$\Delta H = +181$ kJ/mol

Step 1: State the given and needed quantities.

Given:

1.65 grams of NO

$\Delta H = +181$ kJ/mol

Solve:

Heat absorbed in kJ

$N_2\,(g) + O_2\,(g) \rightarrow 2\ NO\,(g)$ $\Delta H = +181$ kJ/mol

1.65 g ? kJ

Step 2: Write an expression for the heat of the reaction and the molar mass needed.

grams of NO → moles of NO → kilojoules of energy

molar mass *heat of reaction*

Step 3: Write conversion factors, including heat of the reaction.

1 mole of NO = 30.01 g of NO:

30.01 g NO / 1 mol NO or 1 mol NO / 30.01 g NO

ΔH for 2 moles of NO = +181 kJ/mol

therefore:

2 mol NO / +181 kJ/mol or +181 kJ/mol / 2 mol NO

Step 4: Set expression to calculate heat.

1.65 g NO × (1 mol NO/30.01 g NO) × (+181 kJ/mol/2 mol NO)

1.65 ~~g NO~~ × (1 ~~mol NO~~/30.01 ~~g NO~~) × (+181 kJ/mol/2 ~~mol NO~~)

1.65 × (1/30.01) × (+181 kJ/mol/2) = 4.98 kJ/mol

4.98 kJ/mol of nitrogen gas is absorbed when 1.65 g of nitrogen monoxide is produced from nitrogen and oxygen.

Quantity relationships in chemical reactions

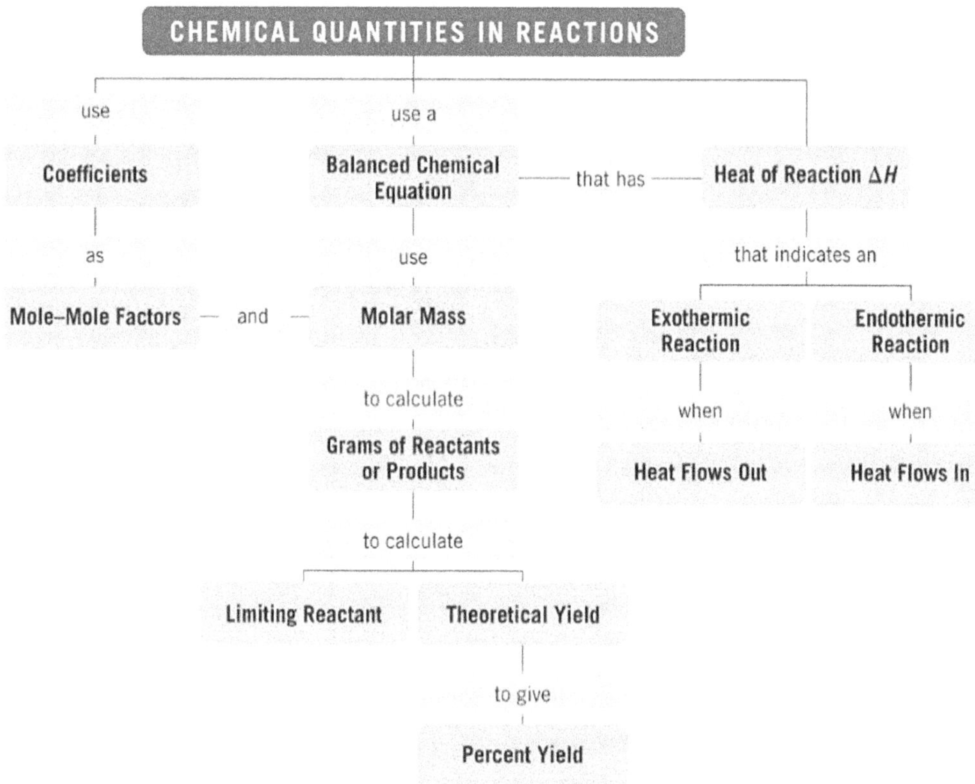

Interrelationship of chemical processes and quantitative outcomes

Thermochemical systems to calculate enthalpy

In addition to Hess' Law of Heat Summation, alternative methods can calculate enthalpy.

For example, if a series of reactions whose enthalpies are known to create an overall reaction that gives the reaction sought, the sum of those enthalpies is the enthalpy of the entire reaction.

Calculating enthalpy change using thermochemical systems of equations is as follows.

For example, from Hess' Law, determine the enthalpy for the reaction:

$$2\ N_2(g) + 5\ O_2(g) \longrightarrow 2\ N_2O_5(g) \qquad \Delta H = ?$$

Use the following heats of reaction:

Equation (1) $2\ H_2(g) + O_2(g) \longrightarrow 2\ H_2O\ (l)$ $\quad \Delta H = -571.7$ kJ

Equation (2) $N_2O_5(g) + H_2O\ (l) \longrightarrow 2\ HNO_3(l)$ $\quad \Delta H = -92$ kJ

Equation (3) $N_2(g) + 3\ O_2(g) \longrightarrow 2\ HNO_3(l)$ $\quad \Delta H = -348.2$ kJ

Manipulate the above equations so that the sum of the desired equation is obtained. Inspect the compounds in each equation and choose an equation with a compound appearing once.

Oxygen (O_2) is in equations (1) and (3), so it is *not* suitable to begin solving the problem.

Dinitrogen pentoxide (N_2O_5) is in equation (2), so equation (2) could begin solving the problem.

Nitrogen is included in equation (3), so equation (3) can be used to begin solving the problem.

For example, solve the problem using nitrogen (N_2) and equation (3). Equation (3) contains one mole of nitrogen as a reactant, and two moles of nitrogen are needed.

Since N_2 is a reactant in both, the equation is not reversed, but equation (3) must be multiplied by two for the required two moles of nitrogen (N_2).

The method used to manipulate the chemical equation is also applied to the heat of the reaction.

Heat of reaction is multiplied by two:

2 × Equation (3):

$2\ N_2(g) + 6\ O_2(g) + 2\ H_2O\ (g) \rightarrow 4\ HNO_3\ (l)$

$\Delta H = \mathbf{2}(-348.7\ \text{kJ})$

$\Delta H = -696.4\ \text{kJ}$

Dinitrogen pentoxide (N_2O_5) is introduced since only equation (2) contains (N_2O_5).

In equation (2), N_2O_5 appears as one mole of reactant.

Reverse the equation and multiply by two.

The sign of the heat of reaction is reversed and doubled:

−2 × Equation (2):

$4\ HNO_3\ (l) \longrightarrow 2\ N_2O_5\ (g) + 2\ H_2O\ (l)$

$\Delta H = -\mathbf{2}(-92\ \text{kJ})$

$\Delta H = +184\ \text{kJ}$

Four moles of nitric acid cancel, which is convenient since it is not in the equation sought.

2 × Equation (3):

$2\ N_2(g) + 6\ O_2(g) + 2\ H_2O\ (g) \longrightarrow \cancel{4\ HNO_3\ (l)}$

$\Delta H = \mathbf{2}(-348.7\ \text{kJ})$

$\Delta H = -696.4\ \text{kJ}$

−2 × Equation (2):

~~$4\ HNO_3\ (l)$~~ $\longrightarrow$ $2\ N_2O_5\ (g) + 2\ H_2O\ (l)$

$\Delta H = -\mathbf{2}(-92\ kJ)$

$\Delta H = +184\ kJ$

Cancel two moles of H_2 and H_2O and one mole of O_2 by reversing equation (1).

Reverse the sign for the heat of reaction:

2 × Equation (3):

$2\ N_2\ (g)$ + ~~6~~ $O_2\ (g) + 2\ H_2O\ (g) \longrightarrow$ ~~$4\ HNO_3\ (l)$~~

$\Delta H = \mathbf{2}(-348.7\ kJ)$

$\Delta H = -696.4\ kJ$

−2 × Equation (2):

~~$4\ HNO_3\ (l)$~~ $\longrightarrow$ $2\ N_2O_5\ (g) + 2\ H_2O\ (l)$

$\Delta H = -\mathbf{2}(-92\ kJ)$

$\Delta H = +184\ kJ$

−1 × Equation (1):

$2\ H_2O\ (\mathrm{l}) \longrightarrow 2\ H_2\ (\mathrm{g}) + O_2\ (\mathrm{g})$

$\Delta H = -\mathbf{1}(571.7\ kJ)$

$\Delta H = +571.1\ kJ$

Summing the chemical equations gives the desired reaction.

Summing the heat of reactions gives the heat of the reaction for the *overall process*:

$2\ N_2\ (g) + 5\ O_2\ (g) \longrightarrow 2\ N_2O_5\ (g)$

$\Delta H = +59\ kJ$

ΔH values used in calculations need not be for actual, measurable processes.

Values from the table of heats of formation have been calculated *via* Hess' Law and can be used for further calculations.

Notes for active learning

Bond Energy and Heat of Formation Relationship

Bond-dissociation energy

Bond dissociation energy is the standard enthalpy change when a bond is broken, and it measures a bond's strength.

Bond energy refers to the heat required to break chemical bonds of *one mole of gaseous molecules* to give *separate gaseous atoms*. Bond energy depends on the types of atoms bonding.

Since energy *input* is required to pull two atoms apart, bond-dissociation energy and bond energy are positive quantities associated with an *endothermic* process. (i.e., absorb energy)

Bond dissociation is the reverse process of *bond formation*; energy is negative and associated with an exothermic process.

For example, the H–H single bond has a bond energy of 436 kJ per mole, but the H–O single bond has a bond energy of 464 kJ per mole.

Common bond energies

Bond	Bond Energy (kJ/mole)	Bond	Bond Energy (kJ/mole)
H–H	436	N–N	159
H–C	414	O–O	138
H–N	389	Cl–Cl	243
H–O	464	C=O	803
H–F	569	N=O	631
H–Cl	431	O=O	498
C–O	351	C≡C	837
C–C	347	N≡N	946

The bond energies in the above chart can be used to estimate $\Delta H°_f$

Bond energy for calculating the heat of formation

For example, calculate ΔH°_f for CH_3OH (*g*) in the following reaction.

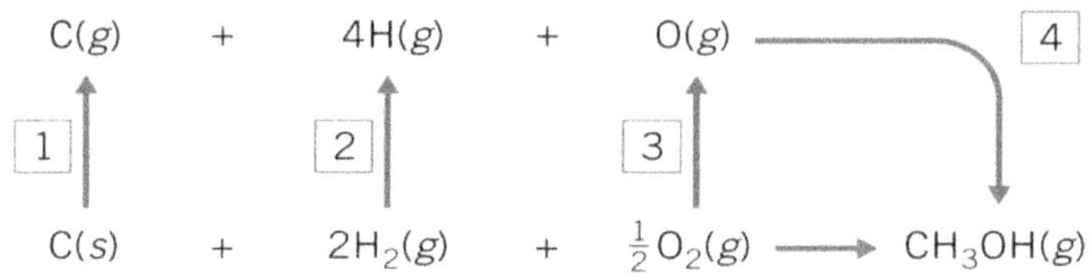

C (*g*)	+	4 H (*g*)	+	O (*g*)		Step 4
Step 1 ↑		Step 2 ↑		Step 3 ↑		↓
C (*s*)	+	2 H_2 (*g*)	+	½ O_2 (*g*)	→	CH_3OH (*g*)

Step 1: **break C–C bonds** (positive energy)

Step 2: **break H–H bonds** (positive energy)

Step 3: **break O–O bond** (positive energy)

Step 4: **form four C–H bonds** (negative energy)

$$\Delta H^\circ_f CH_3OH\ (g) = \Delta H^\circ_f C\ (g) + 4 \times \Delta H^\circ_f H\ (g) + \Delta H^\circ_f O\ (g) - \Delta H_{atom}\ CH_3OH\ (g)$$

$$\Delta H^\circ_f CH_3OH\ (g) = [716.7 + (4 \times 217.9) + 249.2]\ \text{kJ} - \Delta H_{atom}\ CH_3OH\ (g)$$

$$\Delta H^\circ_f CH_3OH\ (g) = 1{,}837.5\ \text{kJ} - \Delta H_{atom}\ CH_3OH\ (g)$$

$$\Delta H^\circ_f CH_3OH\ (g) = 1{,}837.5\ \text{kJ} - 3D_{C—H} + D_{C—O} + D_{O—H}$$

$$\Delta H^\circ_f CH_3OH\ (g) = 1{,}837.5\ \text{kJ} - (3 \times 412) + 360 + 463$$

$$\Delta H^\circ_f CH_3OH\ (g) = 1{,}837.5\ \text{kJ} - 2{,}059\ \text{kJ}$$

$$\Delta H^\circ_f CH_3OH\ (g) = -222\ \text{kJ}$$

ΔH°_f of CH_3OH (*g*) is calculated as –222 kJ using bond energies.

According to the chart, ΔH°_f of CH_3OH (*g*) was experimentally determined to be –201 kJ/mol; estimated bond energies are within 10% (–222 kJ/mol *vs*. –201 kJ/mol).

Gibbs Free Energy (*G*)

Reaction spontaneity

In 1870, Josiah Willard Gibbs (1839-1903) expressed *Gibbs free energy* (G) as the maximum amount of energy present in molecules available *to do non-expansion work* at constant temperature and pressure.

Free energy change (ΔG) is the difference in free energy between the product and reactant molecules.

Gibbs free energy ΔG equation:

$$\Delta G = \Delta H - T\Delta S$$

where G is Gibbs free energy, measured in Joules (J).

Calculating Gibbs free energy

Because ΔG depends on signs, it can be calculated from estimated values.

If ΔG is equal to *zero*, the reaction is at *equilibrium*.

If ΔG is *negative*, the reaction is *spontaneous*.

If ΔG is positive, the reaction is *nonspontaneous*.

For example, if ΔG is *negative* at a given temperature and pressure, the reaction is *spontaneous*.

Chemical bonds may break when molecules react, and new bonds form in the products. Breaking bonds requires energy, while forming bonds releases energy; free energy is exchanged during a reaction. Not all chemical bonds have the same strength or require the same energy.

Gibbs free energy change (ΔG) equals the maximum energy produced during a reaction harnessed as work.

Spontaneous *vs.* nonspontaneous reactions

Table approximates ΔG and indicates whether the reaction is spontaneous or nonspontaneous.

ΔH	ΔS	Result
-	+	Spontaneous at ***all*** temperatures
+	+	Spontaneous at ***high*** temperatures
-	-	Spontaneous at ***low*** temperatures
+	-	Not spontaneous at ***any*** temperature

Reaction energy diagrams

A chemical reaction's thermodynamics can be represented on a *reaction energy diagram* showing the energy of the reactants, products, activation energy, and free energy change ΔG.

Energy appears on the y-axis, and the reaction progress (time) from the starting reactants to the final products appears on the x-axis.

Each reaction has an *energy of activation* (E_{ac}) or *energy barrier* before the reactants transform into products.

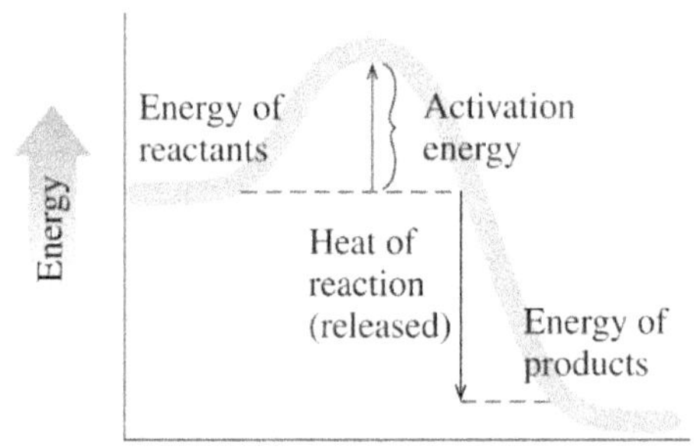

Reaction diagram of energy vs. time (reaction progress)

Gibbs free energy equation

There are several methods to calculate Gibbs free energy ΔG.

Gibbs free energy is a *state function*, so one method is analogous to methods for enthalpy and entropy.

Gibbs free energy expression of $\Delta G°_f$ products *minus* $\Delta G°_f$ reactants:

$$\Delta G°_{rxn} = \Sigma\, n(\Delta G°_f \text{ of products}) - \Sigma\, n(\Delta G°_f \text{ of reactants})$$

where n represents moles of each product or reactant given by the balanced chemical equation coefficients.

Like enthalpy, the standard Gibbs free energy of formation ($\Delta G°_f$) of elements in their standard states is 0 kJ/mol.

Notice the similarity of Gibbs free energy to *Hess' Law of constant heat summation*.

Spontaneous Reactions and ΔG

ΔG for forward and reverse reactions

Gibbs free energy (ΔG) determines whether a reaction is spontaneous or nonspontaneous.

$\Delta G < 0$, the reaction is spontaneous in the *forward* direction

$\Delta G = 0$, the reaction is at *equilibrium* (i.e., rate of forward = rate of reverse reaction)

$\Delta G > 0$, the reaction is spontaneous in the *reverse* direction

ΔG for spontaneous and nonspontaneous reactions

Spontaneous describes a process occurring independently without the input of energy from outside the system, related to the Second Law of Thermodynamics.

ΔG is negative; the reaction is *spontaneous* (or exergonic).

ΔG is positive; the reaction is *nonspontaneous* (or endergonic).

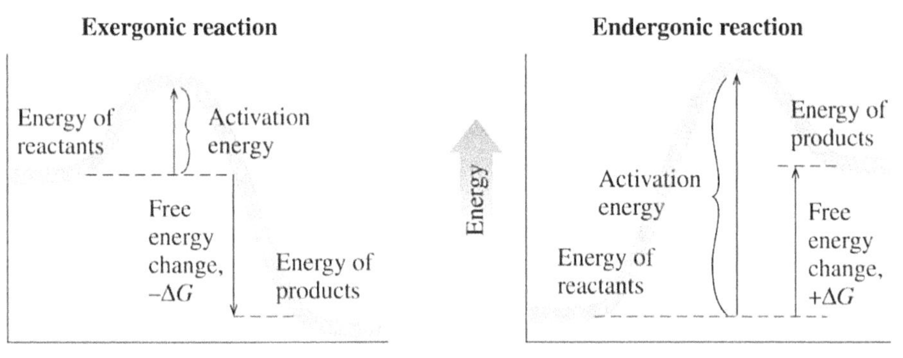

Entropy measures disorder

Entropy (S) is a thermodynamic quantity measuring the spread of a system's energy or disorder.

Increase in entropy of a system corresponds to an increase in system's randomness or disorder.

$\Delta S° < 0$ represents a decrease in entropy

$\Delta S° > 0$ represents an increase in entropy

For example, forming sugar crystals in a supersaturated sugar water solution *increases entropy*.

Entropy increases because the crystals form spontaneously, leading to an increase in entropy.

Gibbs free energy relates enthalpy, temperature, and disorder

Enthalpy, temperature, and entropy relate free energy by:

$$\Delta G^\circ = \Delta H^\circ - T\Delta S^\circ$$

where ΔG° is the change in free energy, ΔH° is the change in enthalpy, T is temperature, and ΔS° is the change in entropy

Exothermic reactions may *not* be spontaneous because of a large, negative ΔS.

Do *not* assume an endothermic reaction is not spontaneous because a large, *positive* ΔS can make it *spontaneous*.

ΔH	ΔS	Sign of ΔG	Spontaneous
–	+	$\Delta G = (-) - [T(+)] = -$	Always, regardless of T
+	–	$\Delta G = (+) - [T(-)] = +$	Never, regardless of T
+	+	$\Delta G = (+) - [T(+)] = ?$	Depends; spontaneous at high T, $-\Delta G$
–	–	$\Delta G = (-) - [T(-)] = ?$	Depends; spontaneous at low T, $-\Delta G$

Table displays how ΔH, ΔS, and T affect spontaneity.

Temperature-controlled reactions

Temperature-controlled reactions are spontaneous at some temperatures but not at others, as indicated in the pie chart's bottom left and top right quadrants.

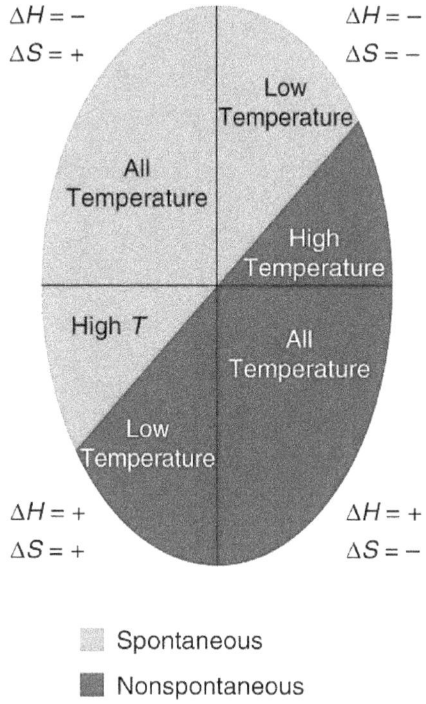

Temperature-controlled reactions

For example, if the enthalpy change is –1,876 kJ, the entropy change is –3.375 kJ/K, and the temperature is 298 K (standard temperature), is the reaction spontaneous? (Units must be consistent.)

Entropy is often expressed in units of J/K and converted to kJ/K.

Substituting these quantities into the equation for *Gibbs free energy*:

$$\Delta G^\circ = \Delta H - T\Delta S$$

$$\Delta G^\circ = -1876 \text{ kJ} - [(298 \text{ K})\cdot(-3.375 \text{ kJ/K})]$$

$$\Delta G^\circ = -870.2 \text{ kJ}$$

Since ΔG° is *negative*, the reaction is *spontaneous*.

Calculating $\Delta G°$ by chemical equilibrium

Use chemical equilibrium to calculate $\Delta G°$:

$$\Delta G° = -RT \ln K$$

$$\Delta G° = -2.303RT \log K$$

where R is the ideal gas law constant (R = 8.314 J/mol·K), T is the temperature in Kelvin, and K is the equilibrium constant

ΔG implies the *direction* and *extent* of a chemical reaction; it does NOT imply a *rate* for the reaction.

Spontaneous reactions may not occur quickly, as they can take years, depending on the kinetics (i.e., rate).

Calculating $\Delta G°$ by electrochemistry

Using electrochemistry to calculate $\Delta G°$:

$$\Delta G° = -nF\text{E}°$$

where n represents the moles of electrons transferred, F is Faraday's constant (96,485 C/mol), and E° is the standard potential for an electrochemical cell

Calculations of Gibbs free energy involve substituting appropriate quantities into expressions.

Thermal Expansion

Thermal expansion of materials

Thermal expansion is the tendency of matter to *change its volume in response to changes in temperature*. Most materials expand when the temperature rises and contract when it falls.

Thermal expansion occurs through *heat transfer*.

When a material changes temperature, it shrinks or expands, depending on the magnitude of the temperature difference.

As temperature increases, molecules within the material gain energy and vibrate at higher rates. Vibration increases the distance between molecules, causing the material to expand.

If the material does not undergo a phase change (e.g., liquid to gas), the expansion of the material can be attributed to the change in temperature.

Coefficient of thermal expansion (α)

Thermal expansion coefficient is an empirically derived constant that describes the expansion of a given material during a temperature rise and depends on the geometry of the material.

Coefficient of thermal expansion (α) is the degree of expansion divided by the change in temperature.

In solid materials, the magnitude of thermal expansion can be calculated to account for:

linear expansion: $\Delta L / L_0 = \alpha\Delta T$

area expansion: $\Delta A / A_0 = 2\alpha\Delta T$

volumetric expansion: $\Delta V / V_0 = 3\alpha\Delta T$

All three calculations require the thermal coefficient of the material. The value of the coefficient α depends on the type of expansion.

Linear expansion

Coefficient of linear expansion is a measure of the expansion (or contraction) per unit of length of a material that occurs when the temperature is increased or decreased by 1 °C. This term is also known as *expansivity*.

Coefficient of thermal expansion (α) is given as:

$$\alpha_{linear} = \frac{\Delta L}{l_i \cdot \Delta T}$$

where α_{linear} is the coefficient of thermal expansion (K^{-1}), ΔL is the change in length (m), l_i is the initial length (m), and ΔT is the change in temperature (K)

Thermal expansion equation for linear expansion is:

$$\Delta L = \alpha L_0 \Delta T$$

where ΔL is the change in length (m), L_0 is the initial length (m), ΔT is the change in temperature (K), and α is the coefficient of linear expansion(K^{-1})

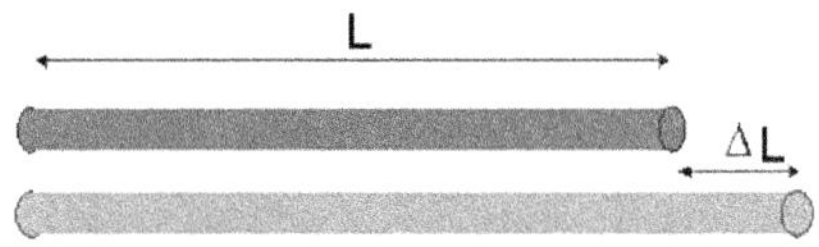

A bar of material is initially at a specific temperature and has a length denoted as L. If the bar is heated such that its temperature rises, it expands linearly by some length ΔL.

For example, a metal rod of 7.00 m is heated to 35 °C. If the rod's length expands to 7.12 m after some time, calculate the linear expansion coefficient (Use room temperature = 27 °C).

Calculate the change in length, ΔL, from the initial L_0 = 7.00 m and expanded length L = 7.12 m.

$$\Delta L = 7.12\text{ m} - 7.00\text{ m}$$

$$\Delta L = 0.12\text{ m}$$

calculate ΔT from initial room temperature T_0 = 27 °C and the heated T = 35 °C.

$$\Delta T = 35\text{ °C} - 27\text{ °C}$$

$$\Delta T = 8\text{ °C}$$

convert ΔT to Kelvin

$$\Delta T = 8 + 273$$

$$\Delta T = 281\text{ K}$$

linear expansion expression:

$$\Delta L / L_0 = \alpha \Delta T$$

isolate α

$$\alpha = \Delta L / (L_0 \times \Delta T)$$

$$\alpha = 0.12\text{ m} / (7\text{ m} \times 281\text{ K})$$

$$\alpha = 6.1 \times 10^{-5}$$

Solution: Coefficient of linear expansion α for the material is 6.1×10^{-5}.

Area expansion

Objects can undergo area expansion due to the same principles.

Thermal expansion equation for area expansion is:

$$\Delta A = \gamma A_0 \Delta T$$

where ΔA is the change in the area (m^2), and γ is the coefficient of area expansion (K^{-1}).

Coefficient of area expansion is expressed as:

$$\alpha_{area} = \frac{\Delta A}{A_i \cdot \Delta T}$$

where α_{area} is the coefficient of thermal expansion (K^{-1}), ΔA is the change in the area (m^2), A_i is the initial area (m^2), and ΔT is the change in temperature (K)

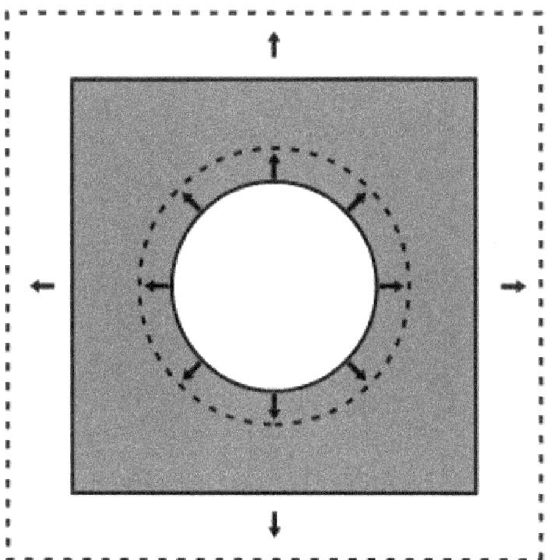

Volume expansion

In general, when a substance is heated, its molecules increase in kinetic energy and move more; thus, a greater average separation is maintained, resulting in increased volume (expansion).

Volume expansion:

$$\Delta V = \beta V_0 \Delta T$$

where ΔV is the volume change (m^3), and β is the coefficient of volume expansion (K^{-1}).

The *coefficient of volume expansion* is expressed as:

$$\alpha_{volume} = \frac{\Delta V}{V_i \cdot \Delta T}$$

where α_{volume} is the coefficient of thermal expansion (K^{-1}), ΔV is the change in volume (m^3), V_i is the initial volume (m^3), and ΔT is the change in temperature (K).

Notes for active learning

Kinetic Theory of Gases

Macroscopic properties of gas

Thermal energy (or *internal energy*) is the sum of the *kinetic energies* of the particles. The thermal energy of a gas is directly proportional to the molecules' kinetic energy (KE).

The *higher* the kinetic energy of the molecules, the more thermal energy the system possesses.

The *lower* the kinetic energy of the molecules, the less thermal energy the system possesses.

Figure below illustrates a closed container of gas where the gas molecules are in random motion throughout the container.

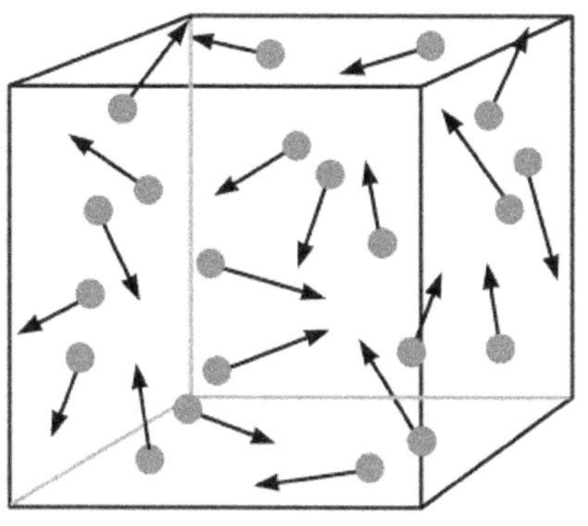

Gas molecules exhibit kinetic energy from random motion in the container

Kinetic Molecular Theory

Kinetic Molecular Theory of Gases relates the microscopic values of atomic kinetic energy to quantities such as temperature and pressure.

Kinetic Molecular Theory (KMT) uses atomic theory to describe the behavior of gases. This theory developed over 100 years, culminating in a 1857 paper by German physicist Rudolf Clausius (1822-1888).

Kinetic Molecular Theory explains the macroscopic properties of a gas (e.g., pressure and temperature) in terms of its microscopic components (e.g., molecules or atoms). It originated in the ancient idea that matter consists of tiny invisible atoms in rapid motion. The idea was revived and developed between the 17th and 19th centuries to explain gas properties.

Five assumptions for gases

Five premises for the kinetic theory of gases:

1. Gases are composed of molecules. Molecules may be treated as perfect spheres of mass (*point masses*), and the space between each of these masses is many times greater than the diameter of the gas molecules.

2. Motion of each gas molecule is *random*, with no order or pattern to either the direction or the magnitude of the velocity.

3. Molecules follow *Newton's Laws of Motion*. Each gas molecule moves in a straight line with a constant velocity. If gas molecules collide, they exert an equal but opposite force on each other.

4. Collisions of molecules are *perfectly elastic*; they *lose no kinetic energy*.

   Therefore, since intermolecular attractive and repulsive forces are nonexistent, the gas molecules' total kinetic energy (KE) remains *constant* before and after the collision.

5. Average kinetic energy (*KE*) for gas particles is proportional to the absolute temperature, regardless of the chemical identity or atomic mass.

   At 0 K (absolute zero or –273 °C), the molecules (and orbiting electrons) are not moving and have no volume.

These rules represent gases as ideal substances, providing only an approximate representation of their actual behavior. However, their description is remarkably accurate. From these assumptions, laws are derived to describe the behavior of gases.

Boltzmann's constant

Final assumption (5) for KE is expressed by:

$$\mathrm{KE} = \tfrac{1}{2} mv^2$$

$$\tfrac{1}{2} mv^2 = (3/2)\, k_{\mathrm{B}}T$$

So

$$\mathrm{KE} = (3/2)\, k_{\mathrm{B}}T$$

where *KE* is kinetic energy, m is mass, v is velocity, T is the temperature (Kelvin), and k_B is *Boltzmann's constant* ($k_{\mathrm{B}} = 1.38 \times 10^{-23}$ J/K).

Boltzmann's constant relates the macroscopic behavior of gases to the microscopic behavior of their particles.

Root-mean-square speed

KE = (3/2) k_BT equation is significant because it states that a gas particle's average kinetic energy is proportional to its absolute temperature. Increasing the temperature of the gas particles increases the total speed and overall energy.

Gas atoms are infinitesimally small; nearly impossible to measure a particle's speed accurately.

Speed of gas particles is defined by the *root-mean-square speed* (u_{rms}) by the equation:

$$u_{rms} = \sqrt{\frac{3RT}{MM}}$$

where R is ideal gas constant, T is absolute temperature (K), and MM is molar mass of gas.

Root-mean-square speed written with the *Boltzmann constant*, k_B:

$$u_{rms} = \sqrt{\frac{3kT}{m}}$$

where k is the Boltzmann constant, T is the absolute temperature (K), and m is the mass of one gas molecule.

Boltzmann distribution plot

Plotting each gas particle's speed distribution at a given temperature shows a slightly asymmetric curve. This speed-distribution curve is the *Maxwell-Boltzmann distribution curve.*

Boltzmann distribution plot shows that the curve's peak corresponds to the most probable speed.

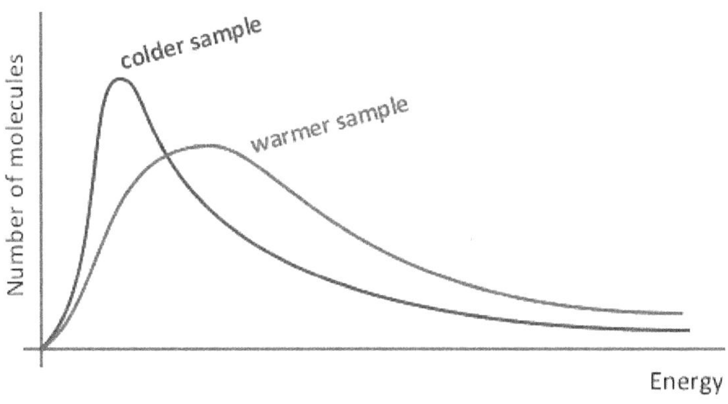

Maxwell-Boltzmann Distribution curves of molecular speeds at two temperatures

If temperature (K) remains constant, the total kinetic energy remains unchanged.

However, the energy is distributed in several ways, and the gas particles travel at many speeds. This distribution changes continually as the gas atoms collide with each other and with the container walls.

The curve flattens (i.e., the net area under the curve is greater) and shifts to the right at higher temperatures, indicating that more gas molecules move at higher speeds and possess more kinetic energy.

Diffusion and effusion

Diffusion is when a substance (solute or particle) spreads from a high-concentration region to a lower concentration. It is explained using the kinetic theory of motion.

According to the kinetic theory, heavier gases diffuse more slowly than lighter gases due to the difference in their molecular speeds.

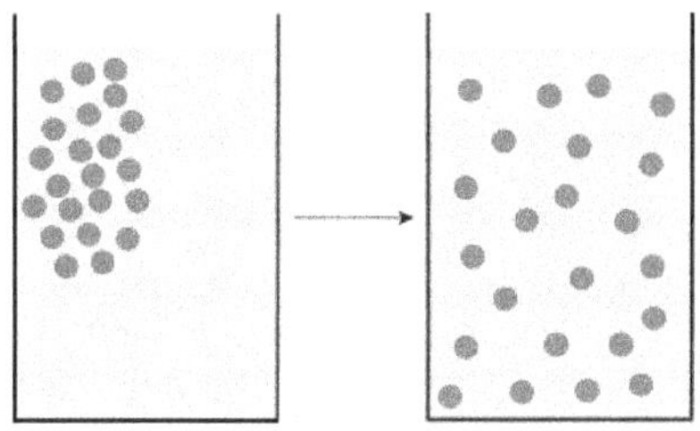

Diffusion of solutes in a solvent

Effusion is the flow of gas particles under pressure from one compartment to another through a small opening, as shown below.

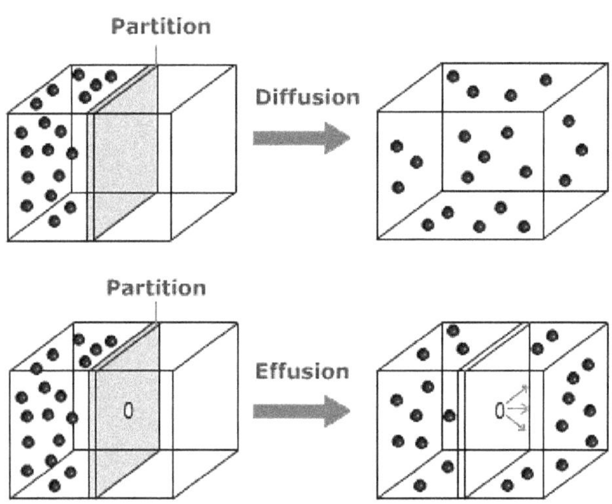

Diffusion and effusion of gas particles

Graham's Law

Scottish chemist Thomas Graham (1805-1869) developed *Graham's Law*, which states that *at constant temperature and pressure conditions, the rates at which two gases diffuse are inversely proportional to the square root of their molar masses.*

Graham's Law is expressed by:

$$\frac{r_1}{r_2} = \sqrt{\frac{MM_2}{MM_1}}$$

where r_1 and r_2 are the diffusion rates of gas 1 and gas 2, respectively, and MM_1 and MM_2 are the molar masses of the gases.

Graham's Law applies to the effusion of gas particles, and the equation for effusion is the same as the equation for diffusion.

Ideal Gas Law

Ideal Gas Law explains the relationship between pressure (P), volume (V), and temperature (T).

Gas law is expressed as:

$$PV = nRT$$

where n is number of moles, and R is universal gas constant with a value of 8.314 J/ mol·K.

One mole is equal to 6.023×10^{23} molecules. Technically, a mole is the number of hydrogen atoms in one gram of hydrogen. Atoms are tiny, so counting them in moles is more practical than counting them individually. Ideal Gas Law applies to the pressure a gas exerts on a cylinder with a moving wall.

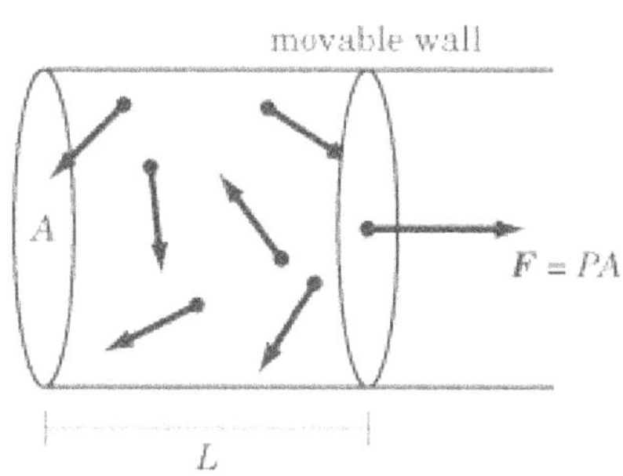

Force that the gas exerts on the wall equals F = PA

Since pressure equals $P = F/A$, the force the gas exerts on the wall equals $F = PA$.

If this force moves the wall back a length of L, the cylinder volume increases by $\Delta V = LA$.

By solving for A and substituting this into the equation for force, the result is:

$$F = P\Delta VL \text{ or } P\Delta V = FL$$

Work is *force multiplied by distance traveled.* By pushing the wall at a distance of L with a force of F, the gas has done work equal to FL.

When gas does work, it symbolizes a change in energy.

If a change in PV is equal to a change in energy, then PV is the total energy of the gas.

For the ideal gas law, nRT is the expression for the total kinetic energy of the gas molecules.

Ideal gas law (PV = nRT) can be related to the number of molecules (N) and Boltzmann's constant (k), which has a value of 1.381×10^{-23} J/K:

$$PV = NkT$$

Other gas laws derive from the Ideal Gas Law by holding specific variables constant.

The number of moles (n) and the gas constant (R) are constant, so four other gas laws result from the Ideal Gas Law:

Boyle's Law

Charles' Law

Combined Gas Law

Closed Container Law

Ideal gases follow KMT (isothermal process)

Kinetic Molecular Theory (KMT) is based on the theoretical concept of an ideal gas.

Ideal gases follow *five* assumptions of KMT, in which gas molecules:

move in *random motion*

are *point masses* with no volume

there are *no intermolecular forces*

collisions are *elastic*

kinetic energy (*KE*) of molecules is proportional to temperature

Compared to an ideal gas, the behavior of a real gas is complicated. By conceptualizing ideal gas behavior, the behavior of real gases becomes more straightforward to comprehend.

In the 17th and 18th centuries, scientists discovered several relationships between the properties of gases. Boyle's, Charles,' and Avogadro's Laws relate pressure, volume, and temperature properties.

These laws were developed empirically and are individual cases of the ideal gas equation:

$$PV = nRT$$

Boyle's Law (isothermal process)

In 1661, Robert Boyle (1627-1691) studied the compressibility of gases. He observed that *the volume of a fixed amount of gas at a given temperature is inversely proportional to the pressure exerted on the gas.*

Boyle's Law applies when the temperature of a gas is held constant. It states that an increase in pressure causes a decrease in volume or that a decrease in pressure causes an increase in volume.

Boyle's Law is expressed as:

$$P_1V_1 = P_2V_2$$

where subscripts 1 and 2 are same gas sample under two sets of pressure and volume conditions.

A plot of volume *vs.* pressure for a gas illustrates that Boyle's Law is a particular case of the ideal gas law, where n (moles of gas) and T (temperature) are constant (figure below).

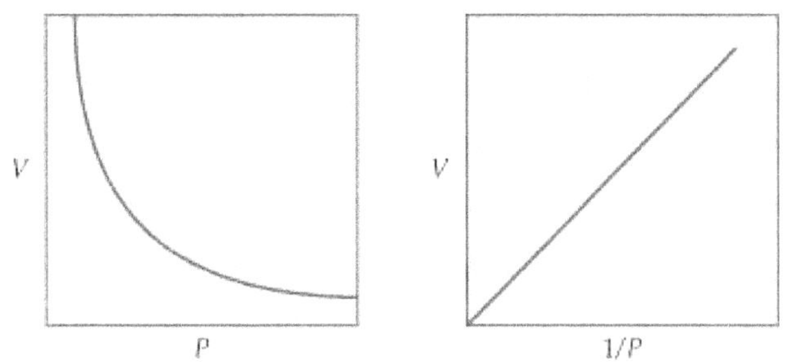

Boyle's Law: as pressure increases, volume decreases

For example, what is the new pressure in a container of gas in a 15.0 L container at a pressure of 5.00 atm when the volume decreases to 0.500 L?

Substitute the known quantities (P_1, V_1, and V_2) into Boyle's Law equation to solve P_2. Note: The volume units must be consistent. In this example, volumes are expressed in units of liters.

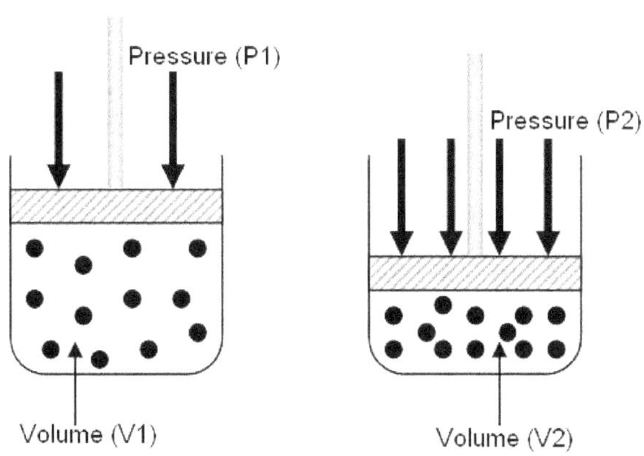

Boyle's Law is where n (moles of gas) and T (temperature) are constant

Boyle's Law:

$$P_1V_1 = P_2V_2$$

Solve for P_2:

$$P_1V_1 / V_2 = P_2$$

$$[(5.00 \text{ atm})\cdot(15.0 \text{ L})] / 0.500 \text{ L} = P_2$$

$$P_2 = 150 \text{ atm}$$

Charles' Law (isobaric process)

French scientist Jacques Charles (1746-1823) developed the relationship between temperature and gas volume.

Charles' Law states that *at constant pressure, the volume of a gas is directly proportional to its absolute temperature (Kelvin)*. It describes the behavior of a gas under constant pressure.

Relationship: volume and temperature are *directly proportional*; that is, as the temperature increases, the volume increases, and as the temperature decreases, the volume decreases.

Charles' Law is expressed as:

$$\frac{V_1}{T_2} = \frac{V_2}{T_2}$$

where subscripts 1 and 2 refer to the same gas sample at different temperature and volume.

Temperature vs. volume plot for a gas illustrates that Charles' Law is another application of the ideal gas law, where n and P are constant, as shown below.

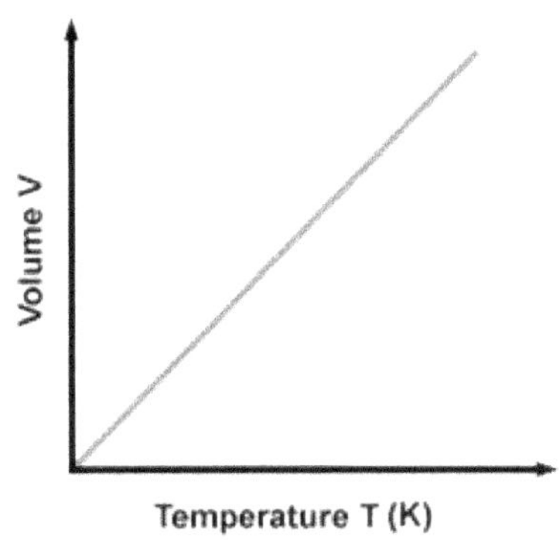

Charles' Law states that as temperature increases, volume increases

For example, if there is 25.0 L of gas at 0 °C, and the temperature is raised to 100 °C, what is the volume of the gas?

First, the temperature is converted to Kelvin.

$T_1 = 0\ °C + 273 = 273\ K$

$T_2 = 100\ °C + 273 = 373\ K$

Substitute the known quantities into the equation for Charles' Law.

$V_1 / T_1 = V_2 / T_2$

Rearrange and solve for V_2:

$V_1\ T_2 / T_1 = V_2$

$(25.0\ L)\cdot(373\ K) / 273\ K = V_2$

$V_2 = 34.2\ L$

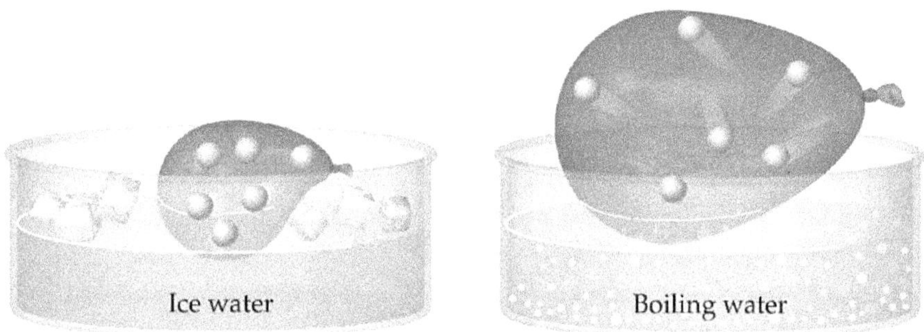

Volume and temperature are directly proportional, when temperature increases, volume increases

Gay-Lussac's Law

In 1802, French chemist Louis Gay-Lussac (1778-1850) proposed *Gay-Lussac's Law,* which states that *the pressure of a given sample of gas is directly proportional at a constant volume to its absolute temperature (Kelvin).*

Gay-Lussac's Law is expressed by:

$$\frac{P_1}{T_1} = \frac{P_2}{T_2}$$

where subscripts 1 and 2 refer to the same gas sample with different temperatures and pressures.

Temperature vs. pressure plot for a gas illustrates that Gay-Lussac's Law is another particular case of the ideal gas law, where n and V are constant.

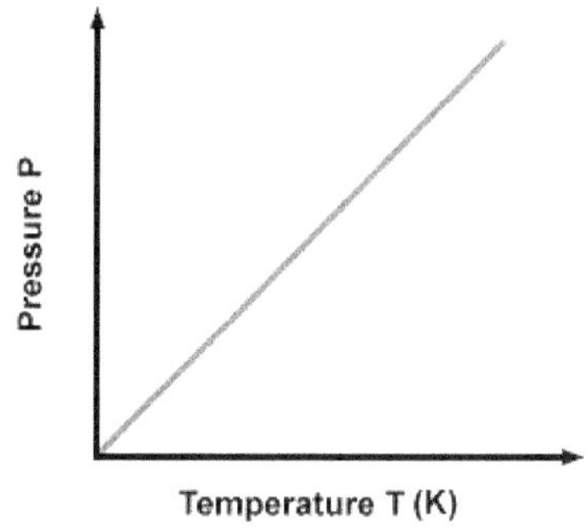

Gay-Lussac's Law states that as temperature increases, pressure increases

For example, suppose a gas is at 30.0 atm pressure and 100 °C, and the temperature changes to 400 °C. What is the new pressure of the gas?

Convert the temperature to Kelvin units.

T_1 = 100 °C + 273 = 373 K

T_2 = 400 °C + 273 = 673 K

Gay-Lussac's Law solves P_2 by substituting the quantities into the equation.

$$\frac{P_1}{T_1} = \frac{P_2}{T_2}$$

P_2 = [(30.0 atm)·(673 K)] / 373 K

P_2 = 54.1 atm

Closed Container Law (constant volume)

Closed Container Law describes the behavior of a gas under constant volume. In such cases, pressure and temperature are directly proportional:

$$\frac{P_i}{T_i} = \frac{P_f}{T_f}$$

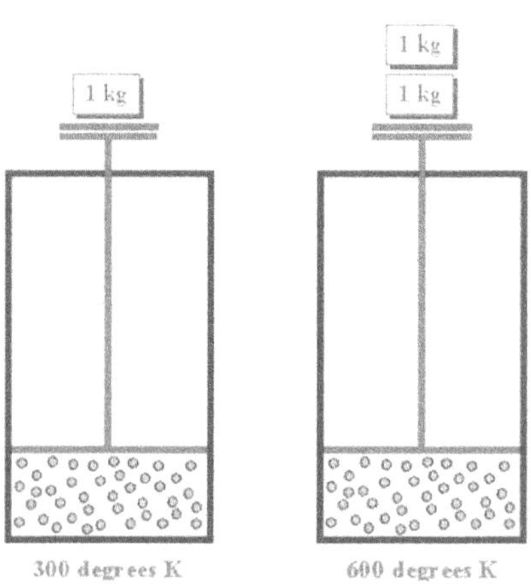

Combined Gas Law

Combined Gas Law combines Boyle's and Charles' Laws. It is *not* the ideal gas law (PV = nRT); it is an equation that relates changes in temperature, volume, and gas pressure.

Combined Gas Law relates the pressure, volume, and temperature of a gas and is expressed as:

$$\frac{P_i V_i}{T_i} = \frac{P_f V_f}{T_f}$$

where subscripts 1 and 2 refer to the same gas sample but under different temperatures, pressures, and volumes.

For example, suppose a gas is at 15.0 atm pressure, with a volume of 25.0 L and a temperature of 300.0 K. What would the volume of the gas be at standard temperature and pressure? (Use standard pressure = 1.00 atm and standard temperature = 273.0 K)

Use the *combined gas law* at standard temperature (T_2) and pressure (P_2):

$(P_1V_1) / T_1 = (P_2V_2) / T_2$ solve for V_2:

$(P_1 \times V_1 \times T_2) / (T_1 \times P_2) = V_2$

$V_2 = [(15.0 \text{ atm})\cdot(25.0 \text{ L})\cdot(273.0 \text{ k})] / [(300.0 \text{ K})\cdot(1.00 \text{ atm})]$

$V_2 = 341 \text{ L}$

Avogadro's Law

Volume of gas in a container is affected by pressure, temperature, and moles of gas.

Relationship between the quantity of gas and its volume was derived by Italian scientist Amadeo Avogadro (1776-1856).

Avogadro's Law states that *gases at a given temperature and pressure occupying a volume are directly proportional to the number of moles of gas present.*

Avogadro's Law is expressed as:

$$\frac{n_1}{V_1} = \frac{n_2}{V_2}$$

where n_1 and n_2 are moles of gas 1 and gas 2, and V_1 and V_2 are the volumes of gas 1 and gas 2.

A plot of *volume vs. moles* of gas (shown below) illustrates that Avogadro's Law is a particular case of the ideal gas law where T and P are held constant.

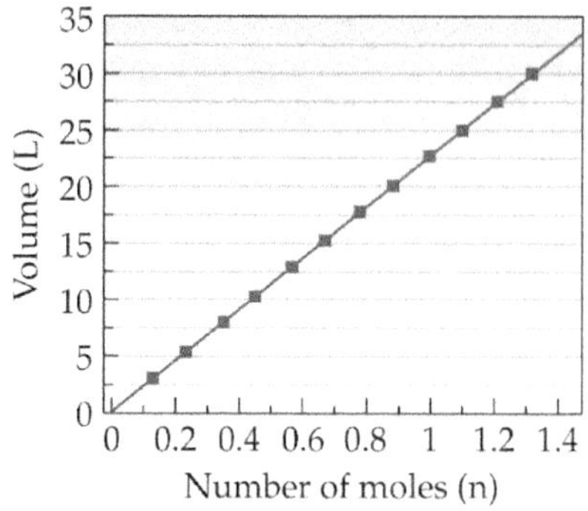

Avogadro's Law states that as moles of a gas increase, volume increases

For example, if 8.00 moles of a gas occupy a volume of 4.00 L at constant pressure and temperature, what volume of gas would 16.0 moles of this gas occupy at the same temperature and pressure?

Use *Avogadro's Law*

$$n_1 / V_1 = n_2 / V_2$$

where n_1 and n_2 are moles of gas 1 and gas 2, and V_1 and V_2 are the volumes of gas 1 and gas 2.

Solve for V_2

$$V_2 = (V_1 \times n_2) / n_1$$

$$V_2 = [(4.00 \text{ L})\cdot(16.0 \text{ mol})] / 8.00 \text{ mol}$$

$$V_2 = 8 \text{ L}$$

Phase Diagrams

Phases of substances

Most substances exist in solid, liquid, or gas form.

> *Solids* have a definite shape and volume (e.g., ice). The molecules in a solid vibrate at a fixed position, and a solid cannot be compressed.
>
> *Liquids* have a definite volume but assume the shape of their container (e.g., water). The molecules move about but are close and bound by intermolecular forces.
>
> *Gases* have neither a definite volume nor a definite shape; they occupy the volume and shape of their container (e.g., steam or water vapor). The molecules move independently and are not held together by intermolecular forces, making a gas easily compressible.

Phases at equilibrium

Phase diagrams indicate the relationship between temperature, pressure, and associated phases, plotting pressure, temperature, or volume for thermodynamically distinct phases (solid, liquid, or gas) at equilibrium.

Phase of a substance (i.e., solid, liquid, or gas) mostly depends on temperature and pressure.

If the pressure is exceptionally high, a substance is less likely to be in a gas phase because the pressure brings the molecules so close that the substance either liquefies or sublimes.

Phase change diagram

When an ice block is placed in the Sun, it slowly absorbs thermal energy through radiation and undergoes phase changes. The ice cube changes from a solid to a liquid to a gas.

> *Melting point* of a substance is the temperature at which it changes from a solid to a liquid, and it has the same temperature range as the *freezing point*; the reverse reaction occurs at the same temperature.
>
> *Boiling point* (or vaporization point) of a substance is the temperature at which it changes phase from a liquid to a gas, within the same temperature range as the *condensation point*.

From the graph below, once the ice cube absorbs enough heat to melt (i.e., reaches 0 °C), its temperature remains constant.

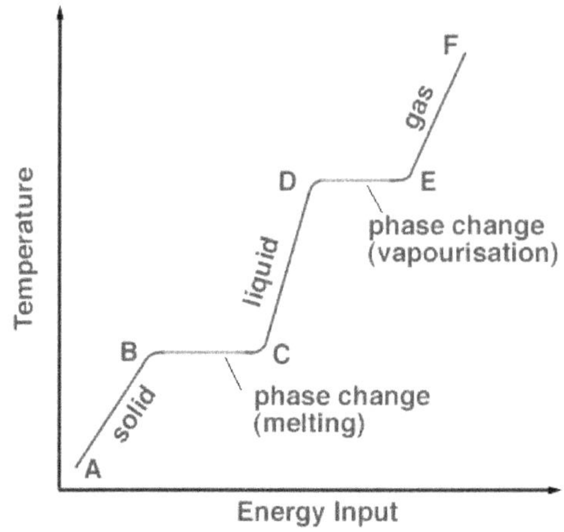

Phase diagram relates temperature changes and states

Melting is points B to C on the graph where all molecules have reached a temperature of 0 °C.

Ice melts only when this occurs, as illustrated by the graph's horizontal plateau from B to C.

When liquid water changes to a gas, the liquid absorbs enough heat to *vaporize* (i.e., at 100 °C).

Liquid temperature remains constant from point D to E until all molecules reach 100 °C.

As illustrated by the *horizontal plateau* from D to E, the *liquid vaporizes* only when this occurs.

Heat during phase changes

Although the energy (heat) input increases with temperature, it does not remain constant, unlike the temperature during a phase change. Technically, heat converts the potential energy stored within each atom into kinetic energy before a phase change occurs.

As energy is continuously added to the process, the substance uses the added heat to transition each molecule *before increasing* the temperature.

Pressure *vs.* temperature relationship

Phase diagram graphically displays the phases at various *pressures* and *temperatures.*

The graph is divided into three unequal parts, representing the substance's three phases.

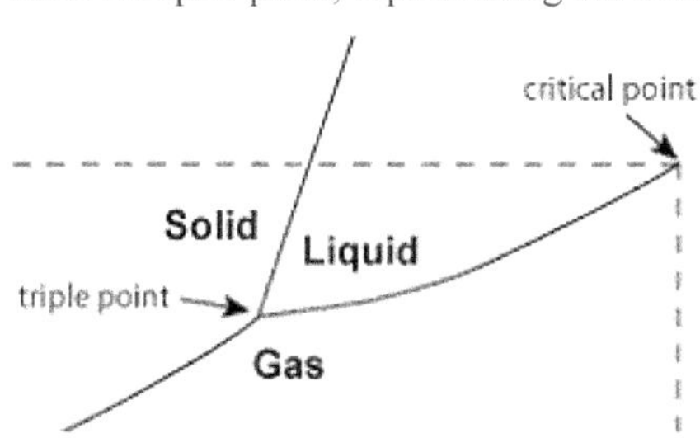

PV phase diagram with triple point and critical point labeled. Dashed horizontal line shows the critical pressure, and the vertical dashed line shows the critical temperature.

Triple point, critical point, and critical temperature

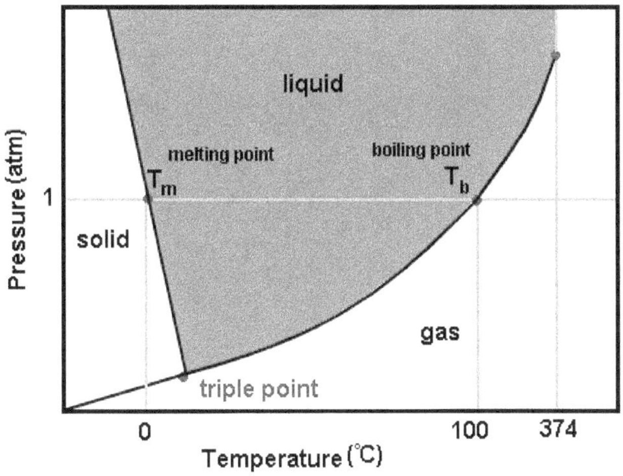

PV phase diagram with triple point and critical point labeled.

Melting points and boiling points are determined by drawing a horizontal line at 1 atm of atmospheric pressure. The horizontal line intersects with a line on the phase diagram, the substance changes phases at that temperature.

Triple point is temperature and pressure at which three phases of matter coexist in equilibrium.

Critical point is temperature and pressure at which liquids and gases become indistinguishable.

Critical temperature is the point at which a *liquid cannot form*, regardless of the pressure.

Phase line boundaries

Solid-liquid boundary is where the *solid and liquid* exist in equilibrium.

Solid-gas boundary is where *solid and gas* exist in equilibrium.

Liquid-gas boundary is where a *liquid* and a *gas* exist in equilibrium.

Positive slope graphs

Solid-liquid phase boundary (or fusion curve) for most substances has a positive slope (i.e., rise over run), so the *melting point increases with pressure*. This is true when the solid phase is denser than the liquid phase.

The greater the pressure, the closer the molecules of the substance are brought together, thereby increasing the effect of intermolecular forces. Thus, the substance requires a higher temperature for its molecules to have enough energy to break the fixed pattern of the solid phase and enter the liquid phase.

A similar concept applies to liquid–gas phase changes.

Negative slope graphs

Water is an exception, which has a solid-liquid boundary with a negative slope so that the melting point decreases with pressure.

Ice (solid water) is less dense than liquid water, as ice floats on water.

At a molecular level, ice is less dense because it has a more extensive network of hydrogen bonding, which requires a greater separation of water molecules.

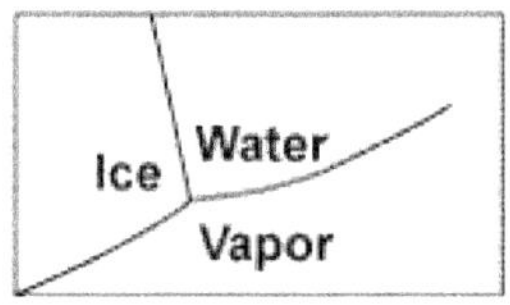

Phase diagram of pressure vs. pressure for water differs because the solid-liquid boundary is slanted to the left. Water (liquid) is denser than ice (solid), if the pressure increases, ice melts into water.

Colligative Properties

Solute *vs.* solvent ratio

Colligative properties depend on the ratio of solute particles to solvent but *not* on the type of chemical species.

Colligative properties include:

1) the relative lowering of vapor pressure,

2) the elevation of boiling point,

3) the depression of freezing point, and

4) osmotic pressure.

Colligative properties calculations incorporate the Van't Hoff factor (*i*), which measures a solute's effect on the solution's colligative properties.

Van't Hoff factor is:

$$\frac{\text{[concentration of particles produced when a substance dissolves]}}{\text{[concentration of a substance calculated from its mass]}}$$

Concentration is expressed as the total number of particles in the solution. (e.g., 5 g of NaCl dissociates into Na^+ and Cl^-).

For example, glucose ($C_6H_{12}O_6$) has an *i* value of 1 because it does not dissociate in solution.

Nonelectrolytes do not dissociate in solutions (e.g., CH_4 and O_2). The salt NaCl has an *i* value of 2 because it dissociates into 1 Na^+ and 1 Cl^- (2 resulting ions per one reactant molecule).

Boiling point elevation ($\Delta T_b = K_b m$)

Adding a solute to a solvent stabilizes the solvent in the liquid phase, thus lowering the solvent molecules' tendency to move to the gas or solid phases. Therefore, the boiling point increases.

Boiling point elevation is proportional to the decrease in vapor pressure in a dilute solution:

$$\Delta T_b = k_b \cdot m \cdot i$$

where ΔT_b is increase in boiling point, k_b is molal boiling point constant (a given value determined experimentally), m is molality (mol solute/kg solvent), and i is Van't Hoff factor.

For example, what is the boiling point elevation if 6.4 g of ammonia is dissolved in 0.3 kg of water? (Use the k_b for water = 0.52 °C/m and the molar mass of ammonia = 17.031 g/mol)

Convert mass of ammonia (NH_3) into moles:

molecular mass = mass / moles

moles = mass / molecular mass

moles = (6.4 g) / (17.031 g/mol)

moles = 0.38 mol NH_3

Calculate molality of the solution:

molality = (moles of solute) / (mass (in kg) of solvent)

m = (0.38 mol NH_3) / (0.3 kg of water)

m = 1.27 m

Calculate boiling point elevation:

$$\Delta T_b = k_b \cdot m \cdot i$$

$$\Delta T_b = (0.52\ °\text{C/m}) \cdot (1.27\ \text{m}) \cdot (1)$$

$$\Delta T_b = 0.66\ °\text{C}$$

Boiling point increased by 0.66 °C due to adding 6.4 g of NH_3.

Freezing point depression ($\Delta T_f = K_f m$)

Solute particles in mixtures increase the strength of intermolecular (e.g., dipole-dipole) bonds. It is more difficult to boil the solution (i.e., boiling point elevation) or freeze it (i.e., freezing point depression).

Freezing point depression (i.e., lowering the freezing point) is calculated by the following equation: (note the negative sign indicates that the change is a decrease).

$$\Delta T_f = -k_f m \cdot i$$

where ΔT_f is the decrease in freezing point, k_f is the molal freezing point constant (a given value determined experimentally), m is the molality (mol solute/kg solvent), and i is the Van't Hoff factor (given since it is determined experimentally).

For example, a 48.0 g sample of a non-electrolyte (does not dissociate in solution) is dissolved in 500.0 g of water to produce a solution with a freezing point of –3.5 °C. What is the molar mass of the compound? (Use k_f of water = 1.86 °C/m, freezing point of pure water = 0 °C)

The freezing point of water = 0 °C, and the freezing point depression, ΔT_f = 3.5 °C.

Calculate the moles using the freezing point depression expression:

$$\Delta T_f = k_f \cdot m \cdot i$$

$$3.5\ °\text{C} = (1.86\ °\text{C/m}) \cdot (x\ /\ 0.5\ \text{kg}) \cdot (1)$$

$$3.5\ °\text{C} = (3.72\ °\text{C/m}) \cdot (x)$$

$$3.5\ °\text{C}\ /\ (3.72\ °\text{C/m}) = x$$

x = 0.94 mol nonelectrolyte

Calculate molar mass by dividing the sample's mass by the number of moles in the sample.

Molar mass = mass / moles

Molar mass = 48.0 g / 0.94 mol

Molar mass = 51.1 g/mol

The molar mass of the unknown nonelectrolyte is 51.1 g/mol.

Notes for active learning

Heat Transformation

Fusion and vaporization

Latent heat (*latent energy* or *heat of transformation*) is the energy released or absorbed by a thermodynamic system during a constant-temperature process — usually a first-order phase transition (i.e., constant temperature during "mixed-phase" states).

Latent heat is the energy supplied or extracted to change the state of a substance without changing its temperature (i.e., first-order phase transition).

Examples include the *latent heat of fusion* (solid to liquid) and the *latent heat of vaporization* (liquid to gas) for phase changes when a substance condenses or vaporizes at a specified temperature and pressure.

Heat and phase changes

Energy is released for:

Condensation: gas changes phase into liquid

Freezing: liquid transforms into a solid

Deposition: gas changes phase directly into a solid

Reverse processes use the same amount of energy.

Vaporization (reverse of condensation): liquid changes phase into a gas

Melting (reverse of freezing): solid changes phase into liquid

Sublimation (reverse of deposition): solid changes phase directly into a gas

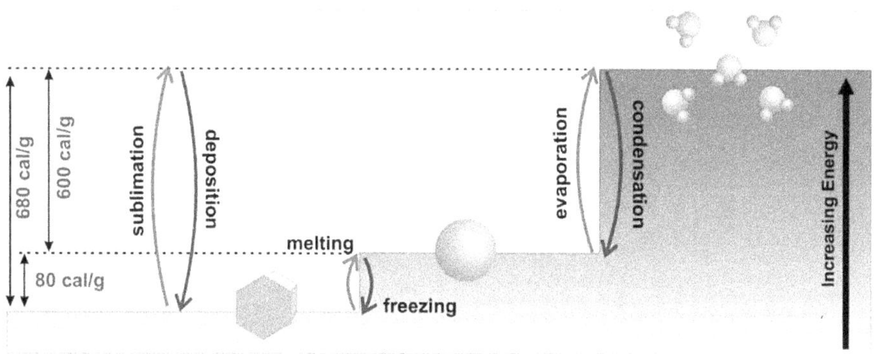

Solid *Liquid* *Vapor*

Latent heat

Latent heat is the amount of heat required for a phase change. Latent heat has been reported for most substances. In general, only computing the total heat is required for the reaction.

Heat required to convert a substance from one phase to another is:

$$Q = m \cdot l$$

where l is the latent heat (J/kg), and m is the mass of the substance (kg).

Latent heats are sometimes given in molar heat or joules per mol (J/mol). Divide by molecular weight of the substance if given the mass in grams.

If any of these processes are reversed, the latent heat is negative (reverse process).

Heat of fusion (ΔH_{fusion})

Heat of fusion (ΔH_{fusion}) is the energy required to melt a given quantity of solid into a liquid at a constant temperature; it is the energy required to melt the solid multiplied by the number of moles.

Heat of fusion ΔH_{fusion} is expressed as:

$$q = n \times \Delta H_{fusion}$$

where q is heat energy, n is the number of moles of the solid, and ΔH_{fusion} is the heat of fusion.

Most substances have reported reference values for the heat of fusion (ΔH_{fusion}).

If the particles become excited (solid to a liquid, liquid to a gas), the change in entropy is always positive because heat is added to the system.

For this process, use latent heat of fusion. The total heat required is the heat of fusion.

If the particles become less excited (gas to a liquid, liquid to solid), the change in entropy is always negative because heat is removed from the system.

A negative change in entropy results in a more ordered molecular structure.

Total heat required for this process is the *heat of vaporization*.

Heat of vaporization (ΔH_{vap})

Heat of vaporization (ΔH_{vap}) is the energy to vaporize a given quantity of liquid to gas at a constant temperature.

It is the energy required to vaporize the liquid times the number of moles of that liquid.

Most substances have reported reference values for the heat of vaporization (ΔH_{vap}).

Heat of vaporization ΔH_{vap} is expressed as:

$$q = n \times \Delta H_{vap}$$

where q is heat energy, and n is number of moles of the liquid, and ΔH_{vap} is heat of vaporization.

Sublimation is when a solid bypasses the liquid phase and transitions directly into the gas phase.

The heat required for *sublimation* sums up the *heat of fusion* and the *heat of vaporization.*

Deposition is the reverse reaction of sublimation.

$$Q_{sub} = Q_{fus} + Q_{vap}$$

Total heat of reaction

For a process with multiple phase changes, the total heat of the reaction is the sum of the separate heat needed for each phase change and the addition of heat.

For these problems, consider the First Law of Thermodynamics, which states that heat must be conserved (heat gained = heat loss), and the equation for heat ($Q = mc\Delta T$), where c is the specific heat capacity, and m is the mass.

For example, when calculating the change in heat (change in entropy) when 2 kg of ice at 0 °C is warmed to 120 °C, several equations are required using the known values.

- melting point = 0 °C
- boiling point = 100 °C
- latent heat of fusion = 3.33×10^5 J/kg
- latent heat of vaporization = 2.26×10^6 J/kg
- specific heat capacity = 4,186 J/kg °C

Heat required to phase change the *ice into liquid water*:

$$Q = m\, l_{\text{fusion}}$$

$$Q = (2\text{kg})\cdot(3.33 \times 10^5\ \text{J/kg}) = 6.66 \times 10^5\ \text{J}$$

Heat required to heat the *water to its boiling point*:

$$Q = \text{mc}\Delta\text{T}$$

$$Q = (2\ \text{kg})\cdot(4186\ \text{J/kg °C})\cdot(100\ \text{°C} - 0\ \text{°C}) = 837{,}200\ \text{J}$$

Heat required to phase change the *liquid into gas:*

$$Q = m\,l_{vap} = (2\text{kg})\cdot(2.26 \times 10^6\ \text{J/kg})$$

$$Q = 4.52 \times 10^6\ \text{J}$$

Heat required to heat the water to 120 °C:

$$Q = mc\Delta\text{T}$$

$$Q = (2\ \text{kg})\cdot(4{,}186\ \text{J/kg °C})\cdot(120\ °\text{C} - 100\ °\text{C}) = 167{,}440\ \text{J}$$

Total heat required for the process to occur is the *sum of the separate heats*:

$$Q_{total} = Q_{fusion} + Q_{heat\_1} + Q_{vap} + Q_{heat\_2}$$

$$Q_{total} = 6.66\times10^5 + 837{,}200 + 4.52\times10^6 + 167{,}440$$

$$Q_{total} = 6{,}190{,}640\ \text{J}$$

Calculating heat of fusion and heat of vaporization

Heat of fusion and heat of vaporization are measured in units of kJ/mol.

However, they may be expressed as J/g, where energy is obtained by multiplying the latent heat by the substance's mass rather than moles.

For example, how many kJ are required when 31.5 g of H_2O melts at its melting point of 0 °C? (Use molar mass for H_2O = 18 g/mol and molar heat of fusion for water ΔH_{fusion} = 6.02 kJ/mol).

Use molar mass relationship:

mass (g) = moles (mol) × molar mass (g/mol)

Rearrange to find the number of moles:

moles (g) = mass (g) / molar mass (g/mol)

moles = (31.5 g) / (18 g/mol)

moles = 1.75 mol

Use heat of fusion (ΔH_{fusion}):

$$q = n \times \Delta H_{fusion}$$

$$q = 1.75\ \text{mol} \times 6.02\ \text{kJ/mol}$$

$$q = 10.54\ \text{kJ}$$

10.54 kJ heat (i.e., positive value) is required to melt 31.5 g of water.

Pressure *vs.* Volume Phase Diagrams

PV graphs

Pressure *vs.* volume diagram (PV graph) describes the changes in system pressure and volume.

PV diagrams show the thermodynamic process by graphing pressure against volume (P *vs.* V).

Work done by the system is equal to the area under the curve.

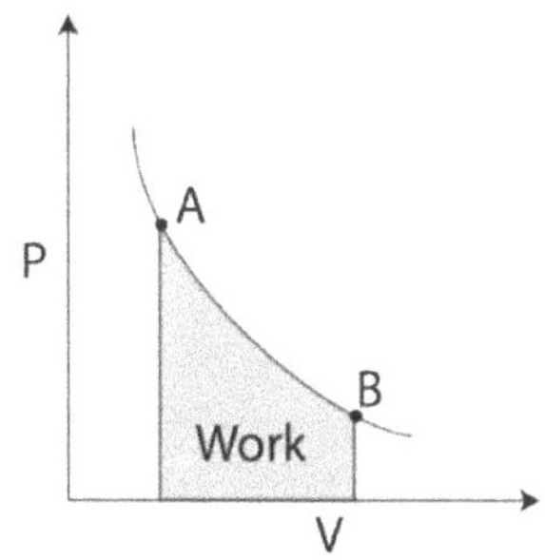

Area under the curve on the above PV diagram equals the work done

Cyclic pressure *vs.* volume reactions

In cyclic reactions, work done is equal to the *area enclosed* by the graph.

These reactions are often comprised of different processes, as shown below.

work is *done **by*** the gas when its volume increases, and its pressure decreases.

work is *done **on*** the gas when its volume decreases and its pressure increases.

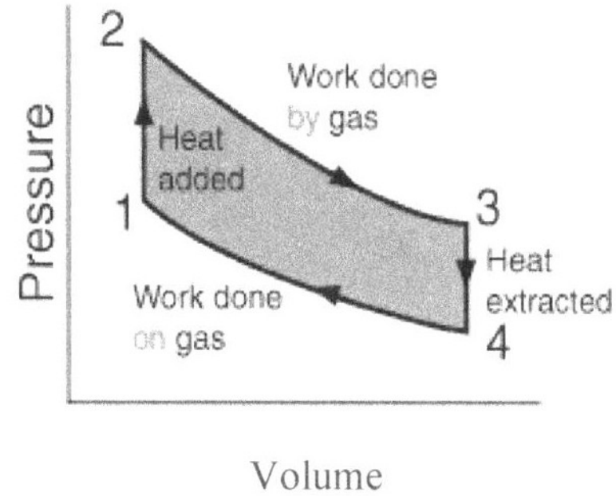

*Points 4 to 1 show work done **on** the gas; points 2 to 3 work done **by** the gas*

Calculating work done by gas

Work can be calculated for the gas phase of a gas and its container.

Gases may *expand* or *compress*; the *pressure* of the gas and *volume changes* determine work:

Figure illustrates:

expansion is work done ***by*** the system ($\Delta E = q - w$)

compression is work done ***on*** the system ($\Delta E = q + w$)

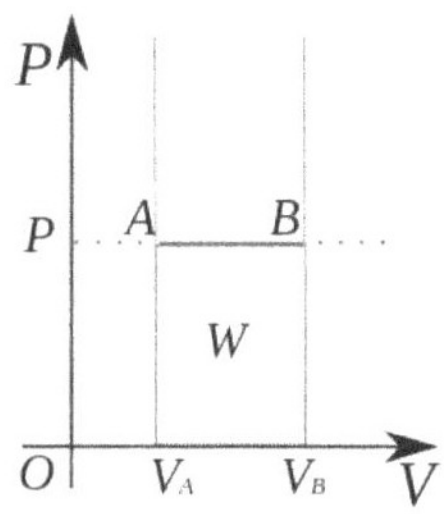

Shaded area (W) represents the work done by the gas

Positive and negative volume changes

Volume change (ΔV) is the final minus initial volume of gas:

$$\Delta V = V_{final} - V_{initial}$$

Volume is increased to *expand* a gas (ΔV is *positive*).

Gas (of the system) must do work (w) on surroundings; thus, work must be *negative*.

Volume is decreased to compress a gas (ΔV is *negative*).

Surroundings must do work (w) on the system; thus, work must be *positive*.

Thermodynamic work cycles

Processes may labeled *adiabatic*, *isothermal*, *isobaric*, and *isochoric.*

Adiabatic process has *no transfer of heat* (Q = 0); the change in energy equals the work done.

Adiabatic processes are often confused with isothermal processes because it is sometimes assumed that no heat transfer means no change in temperature. However, this is not true—in isothermal instances, heat energy must be allowed into (or out of) the reaction to maintain a constant temperature.

Adiabatic processes, unlike isothermal reactions, occur quickly and have a PV graph with the same shape, but are *much steeper* than isothermal processes.

Adiabatic processes

Adiabatic process is when energy transfer within the system results in *work only*.

For example, a rapid expansion or contraction of gas is nearly adiabatic.

Adiabatic processes have *no heat* transfer.

Adiabatic processes *cannot decrease entropy*.

temperature is constant

$q = 0$

$\Delta E = w$

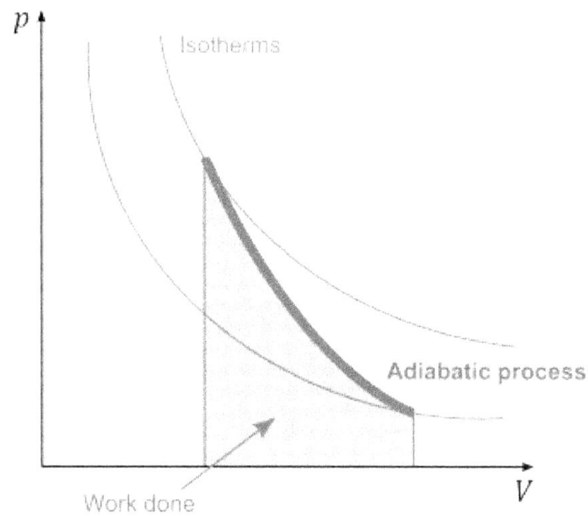

Adiabatic process PV diagrams of work with isothermal lines called isotherms

Adiabatic processes include a tire pump or a sharp expulsion of breath.

Isochoric processes

Isochoric process (or *constant-volume process*, *isovolumetric process*, and *isometric process*) is when the volume of the closed system remains *constant*; these include those inside a closed and rigid container or a constant volume thermometer.

volume is constant

$w = 0$

$\Delta E = q$

A thermodynamic process is the *addition or removal of heat*. Isolation of the container's contents establishes the closed system, and the inability of the container to deform imposes the *constant-volume* condition.

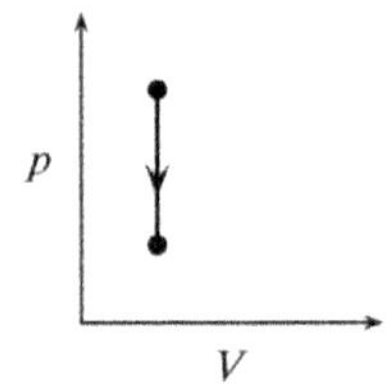

Isochoric process PV diagrams with constant volume show a vertical line

Isochoric processes include heating (or cooling) the contents of a sealed, inelastic container.

Isothermal processes

Isothermal reactions maintain a constant temperature ($\Delta T = 0$). The change in internal energy must stay the same, as an increase in thermal energy (heat) increases temperature ($\Delta U = 0$).

temperature is constant

$\Delta T = 0$

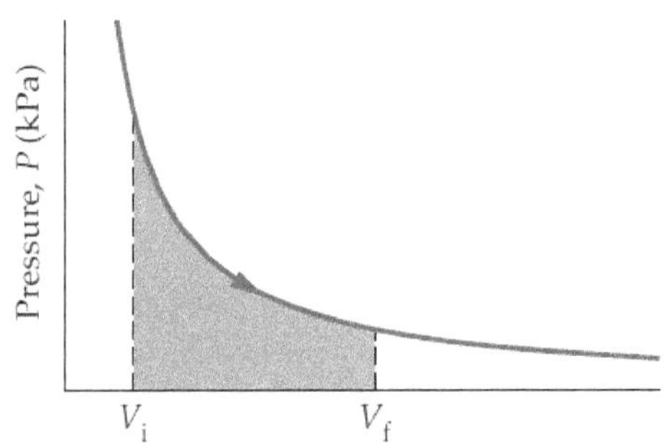

Isothermal process PV diagram at constant temperature

Isothermal process is when the temperature of the system remains constant, typically when a system is in contact with an outside thermal reservoir. A change in the system enables it to adjust the temperature via the reservoir through heat exchange.

Isobaric processes

Isobaric process is when pressure of the system remains constant.

Isobaric reactions maintain constant pressure; work equals pressure times volume change ($W = P\Delta V$).

pressure is constant

$\Delta P = 0$

$w = P\Delta V$

Examples of isobaric reactions include a piston in an engine or a flexible container open to Earth's atmosphere.

When pressure is constant, the PV graph of an isobaric process is a horizontal line.

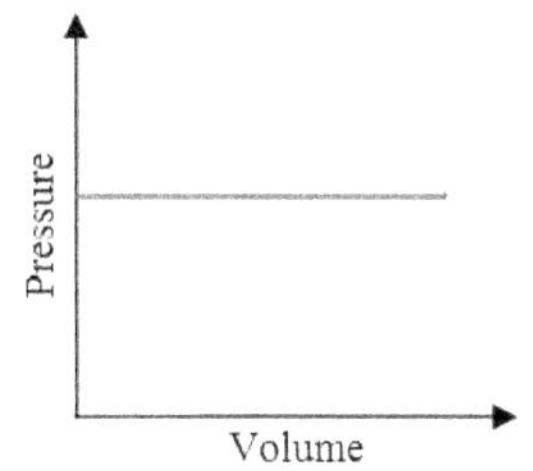

Isochoric process is an isovolumetric process because volume remains constant.

Since there is no movement, the work is zero, and change in energy is due to the amount of heat added to the system.

$W = 0$

$\Delta E = Q$

Heat transferred to the system does work and *changes* the internal energy (U) of the system.

Thermodynamic work processes summary

Process	Constant	PV Diagram	Ideal Gas Law	First Law of Thermodynamics
Isobaric	Pressure	Horizontal line	V α T	$\Delta U = Q + W$
Isochoric	Volume	Vertical line	P α T	$\Delta U = Q$
Isothermal	Temperature	Curved line	PV α T	$\Delta U = 0$
Adiabatic	No heat exchanged	Curved line (jumps to different isotherm)	PV = nRT (only "nR" are constant)	$\Delta U = W$

Thermodynamic work processes, with relationship to Ideal Gas Law ($PV = nRT$)

Summary

Heat and temperature

Heat is the total internal energy due to *molecular motion.*

Temperature is the *average kinetic energy* (KE) for a specific substance.

Calories are the heat needed to raise the temperature of 1 gram of water by 1 ° Celsius (g/°C).

Kelvin (K) is the measure of absolute temperature.

Absolute zero = 0 K

Kelvin and Celsius use the same increments.

Specific heat is the ratio of energy absorbed to temperature rise per unit mass.

Calorimeter measures the heat of a reaction.

Temperature remains *constant* during phase change.

Phase diagram

Triple point: three phases exist simultaneously

Critical point: the substance is always a gas

$PV = NkT$

Boltzmann's constant 1.381×10^{-23} J/K

Heat transfer

Conduction requires *direct contact.*

Rate of heat transfer (H) by conduction:

$$H = \frac{kA\Delta T}{L}$$

Convection uses energized particles from one place to another.

Radiation uses electromagnetic waves

Laws of thermodynamics

Zeroth Law of thermodynamics: two systems in thermal equilibrium with a third system are in thermal equilibrium.

First Law of thermodynamics: energy is always conserved:

$$\Delta U = Q - W$$

Work was done *by* a system (−)

Work was done *on* a system (+)

$Q < 0$ is exothermic, where Q is heat

$Q > 0$ is endothermic, where Q is heat

Second Law of thermodynamics: *entropy* increases or stays constant, never decreases

Entropy (S) is a measure of disorder within a system. The natural tendency is to move towards maximum disorder:

$$\Delta S \geq \frac{Q}{T}$$

$\Delta S_{\text{universe}} > 0$ spontaneous

$\Delta S_{\text{universe}} = 0$ at equilibrium

$\Delta S_{\text{universe}} < 0$ non-spontaneous (reverse process spontaneous)

Third Law of thermodynamics: as the temperature of a system approaches absolute zero, the entropy (S) approaches a constant value (typically zero)

$$S = k \cdot ln\,(\Omega)$$

where S is the entropy (J/K), k is Boltzmann's constant (1.381×10^{-23} J/K), and Ω is the number of microstates

Entropy of a system at absolute zero is zero.

However, some systems have more than one minimum energy state

Enthalpy

Amount of heat used at constant pressure:

$$\Delta U = \Delta H - P\Delta V$$

$\Delta H < 0$ exothermic

$\Delta H > 0$ endothermic

Hess' Law: $\Delta H_{rxn} = H_{products} - H_{reactants}$

Gibb's free energy

$\Delta G = \Delta H - T\Delta S$

$G < 0$: spontaneous reaction

$G = 0$: reaction is at equilibrium

$G > 0$: nonspontaneous reaction

Thermal expansion

Coefficient of linear expansion (or *expansivity*) measures the expansion or contraction per unit of length of a material when temperature increases or decreases by 1 °C under constant pressure.

Expansivity:

$$\alpha_{linear} = \frac{\Delta l}{l_i \cdot \Delta T}$$

where α_{linear} is the coefficient of thermal expansion (K^{-1}), Δl is the change in length (m), l_i is the initial length (m), and ΔT is the change in temperature (K)

Coefficient of area expansion:

$$\alpha_{area} = \frac{\Delta A}{A_i \cdot \Delta T}$$

where α_{area} is the coefficient of thermal expansion (K^{-1}), ΔA is the change in the area (m^2), A_i is the initial area (m^2), and ΔT is the change in temperature (K)

Coefficient of volume expansion:

$$\alpha_{volume} = \frac{\Delta V}{V_i \cdot \Delta T}$$

where α_{volume} is the coefficient of thermal expansion (K^{-1}), ΔV is the change in volume (m^3), V_i is the initial volume (m^3), and ΔT is the change in temperature (K)

Kinetic Theory of Gases

Five critical assumptions for the Kinetic Theory of Gases.

1. Gas molecules move in *random molecular motion* and continue in a straight line until they collide with something.

2. Gas molecules are a "*point mass*," a particle with a mass of nearly zero.

   An ideal gas particle has a *negligible volume.*

3. *No attractive intermolecular* forces for gases participate.

4. *Collisions* between gas molecules are *perfectly elastic.*

   Therefore, the gas molecules' total kinetic energy (KE) remains *constant before and after* the collision since intermolecular attractive and repulsive forces are nonexistent.

5. Average kinetic energy (*KE*) for gas particles is proportional to the absolute temperature, regardless of the chemical identity or atomic mass.

   At 0 K (absolute zero or –273 °C), the molecules (and orbiting electrons) are not moving and have no volume.

Latent heat

Latent heat (l) is the heat (Q) required for a *phase change*:

$$Q = m \cdot l$$

where l is the latent heat (J/kg), and m is the mass (kg)

PV diagrams

Work – *area under pressure vs. volume* curve

Isothermal – *constant temperature*: hyperbola

Adiabatic – *no transfer of heat*: steeper hyperbola

Isobaric – *constant pressure*: horizontal line

Isochoric – *constant volume*: vertical line

Phase diagram: pressure and temperature

Melting point is the temperature at which a substance changes from solid to liquid.

Freezing point is the same temperature as melting point but is the reverse reaction when a substance changes from liquid to solid.

Boiling point (or *vaporization*) is temperature when a substance changes from liquid to gas.

Condensation point is the same temperature as boiling point but is the reverse reaction when a substance changes from gas to liquid.

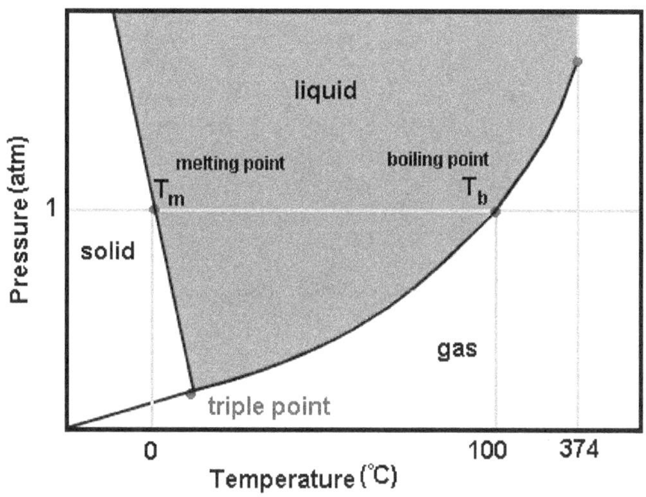

Notes for active learning

PRACTICE QUESTIONS
&
DETAILED EXPLANATIONS

Practice Questions

Practice Set 1: Questions 1–20

1. Which of the following statement regarding the symbol ΔG is NOT true?

A. Specifies the enthalpy of the reaction
B. Refers to the free energy of the reaction
C. Predicts the spontaneity of a reaction
D. Describes the effect of enthalpy and entropy on a reaction
E. None of the above

2. What happens to the kinetic energy of a gas molecule when the gas is heated?

A. Depends on the gas
B. Kinetic energy increases
C. Kinetic energy decreases
D. Kinetic energy remains constant
E. None of the above

3. How much heat energy (in Joules) is required to heat 21.0 g of copper from 21.0 °C to 68.5 °C? (Use the specific heat c of Cu = 0.382 J/g·°C)

A. 462 J **B.** 188 J **C.** 522 J **D.** 662 J **E.** 381 J

4. For n moles of gas, which term expresses the kinetic energy?

A. nPA, where n = number of moles of gas, P = total pressure, and A = surface area of the container walls
B. ½nPA, where n = number of moles of gas, P = total pressure, and A = surface area of the container walls
C. ½MV^2, where M = molar mass of the gas and V = volume of the container
D. MV^2, where M = molar mass of the gas and V = volume of the container
E. 3/2 nRT, where n = number of moles of gas, R = ideal gas constant, and T = absolute temperature

5. Which of the following is NOT an endothermic process?

A. Condensation of water vapor
B. Boiling liquid
C. Water evaporating
D. Ice melting
E. All are endothermic

6. What is true of an endothermic reaction if it causes a decrease in the entropy (S) of the system?

A. Only occurs at low temperatures when ΔS is insignificant
B. Occurs if coupled to an endergonic reaction
C. Never occurs because it decreases ΔS of the system
D. Never occurs because ΔG is positive
E. None of the above

7. Which of the terms describe(s) energy contained in an object or transferred to an object?

I. chemical II. electrical III. heat

A. I only
B. II only
C. I and II only
D. I and III only
E. I, II and III

8. The greatest entropy is observed for which 10 g sample of CO_2?

A. CO_2 (*g*)
B. CO_2 (*aq*)
C. CO_2 (*s*)
D. CO_2 (*l*)
E. All are equivalent

9. What is the term for a reaction that proceeds by absorbing heat energy?

A. Isothermal reaction
B. Exothermic reaction
C. Endothermic reaction
D. Spontaneous
E. None of the above

10. If a chemical reaction has $\Delta H = X$, $\Delta S = Y$, $\Delta G = X - RY$ and occurs at R K, the reaction is:

A. spontaneous
B. at equilibrium
C. nonspontaneous
D. irreversible
E. cannot be determined

11. The thermodynamic systems that have high stability tend to demonstrate:

A. maximum ΔH and maximum ΔS
B. maximum ΔH and minimum ΔS
C. minimum ΔH and maximum ΔS
D. minimum ΔH and minimum ΔS
E. none of the above

12. Whether a reaction is endothermic or exothermic is determined by:

A. an energy balance between bond breaking and bond forming, resulting in a net loss or gain of energy
B. the presence of a catalyst
C. the activation energy
D. the physical state of the reaction system
E. none of the above

13. What role does entropy play in chemical reactions?

A. The entropy change determines whether the reaction occurs spontaneously
B. The entropy change determines whether the chemical reaction is favorable
C. The entropy determines how much product is produced
D. The entropy change determines whether the reaction is exothermic or endothermic
E. The entropy determines how much reactant remains

14. Calculate the value of $\Delta H°$ of reaction using the provided bond energies.

$H_2C{=}CH_2\ (g) + H_2\ (g) \rightarrow H_3C{-}CH_3\ (g)$

C–C: 348 kJ C≡C: 960 kJ

C=C: 612 kJ C–H: 412 kJ H–H: 436 kJ

A. –348 kJ
B. +134 kJ
C. –546 kJ
D. –124 kJ
E. –238 kJ

15. The bond dissociation energy is:

I. useful in estimating the enthalpy change in a reaction

II. the energy required to break a bond between two gaseous atoms

III. the energy released when a bond between two gaseous atoms is broken

A. I only

B. II only

C. I and II only

D. I and III only

E. I, II and III

16. Based on the following reaction, which statement is true?

$N_2 + O_2 \rightarrow 2\ NO$ (Use the value for enthalpy, $\Delta H = 43.3$ kcal)

A. 43.3 kcal are consumed when 2.0 mole of O_2 reacts

B. 43.3 kcal are consumed when 2.0 moles of NO are produced

C. 43.3 kcal are produced when 1.0 g of N_2 reacts

D. 43.3 kcal are consumed when 2.0 g of O_2 reacts

E. 43.3 kcal are produced when 2.0 g of NO are produced

17. Which of the following properties of gas is/are a state function

I. temperature II. heat III. work

A. I only

B. I and II only

C. II and III only

D. I, II and III

E. II only

18. The following statements concerning temperature change as a substance is heated are correct, EXCEPT:

A. As a liquid is heated, its temperature rises until its boiling point is reached

B. During the time a liquid is changing to the gaseous state, the temperature gradually increases until all the liquid is changed

C. As a solid is heated, its temperature rises until its melting point is reached

D. During the time for a solid to melt to a liquid, the temperature remains constant

E. The temperature remains the same during the phase change

19. Calculate the value of $\Delta H°$ of reaction for:

$$O{=}C{=}O\ (g) + 3\ H_2\ (g) \rightarrow CH_3{-}O{-}H\ (g) + H{-}O{-}H\ (g)$$

Use the following bond energies, $\Delta H°$:

C–C: 348 kJ	C=C: 612 kJ	C≡C: 960 kJ	C–H: 412 kJ
C–O: 360 kJ	C=O: 743 kJ	H–H: 436 kJ	H–O: 463 kJ

A. –348 kJ
B. +612 kJ
C. –191 kJ
D. –769 kJ
E. +5,779 kJ

20. Which statement(s) is/are true for ΔS?

I. ΔS of the universe is conserved
II. ΔS of a system is conserved
III. ΔS of the universe increases with each reaction

A. I only
B. II only
C. III only
D. I and II only
E. I and III only

Notes for active learning

Practice Set 2: Questions 21–40

21. Which of the following reaction energies is the most endothermic?

A. 360 kJ/mole
B. –360 kJ/mole
C. 88 kJ/mole
D. –88 kJ/mole
E. 0 kJ/mole

22. A fuel cell contains hydrogen and oxygen gas that react explosively, and the energy converts water to steam which drives a turbine to turn a generator that produces electricity. The fuel cell and the steam represent which forms of energy, respectively?

A. Electrical and heat energy
B. Electrical and chemical energy
C. Chemical and heat energy
D. Chemical and mechanical energy
E. Nuclear and mechanical energy

23. Which of the following is true for the ΔG of formation for $N_2\ (g)$ at 25 °C?

A. 0 kJ/mol
B. positive
C. negative
D. 1 kJ/mol
E. more information is needed

24. Which of the statements best describes the following reaction?

$$HC_2H_3O_2\ (aq) + NaOH\ (aq) \rightarrow NaC_2H_3O_2\ (aq) + H_2O\ (l)$$

A. Acetic acid and NaOH solutions produce sodium acetate and H_2O
B. Aqueous solutions of acetic acid and NaOH produce aqueous sodium acetate and H_2O
C. Acetic acid and NaOH solutions produce sodium acetate solution and H_2O
D. Acetic acid and NaOH produce sodium acetate and H_2O
E. An acid plus a base produce H_2O and a salt

25. If a chemical reaction is spontaneous, which value must be negative?

A. C_p
B. ΔS
C. ΔG
D. ΔH
E. K_{eq}

26. What is the purpose of the hollow walls in a closed hollow-walled container that effectively maintains the temperature inside?

A. To trap air trying to escape from the container, which minimizes convection
B. To act as an effective insulator, which minimizes convection
C. To act as an effective insulator, which minimizes conduction
D. To provide an additional source of heat for the container
E. Reactions occur within the walls that maintain the temperature within the container

27. If the heat of reaction is exothermic, which of the following is always true?

A. The energy of the reactants is greater than the products
B. The energy of the reactants is less than the products
C. The reaction rate is fast
D. The reaction rate is slow
E. The energy of the reactants is equal to that of the products

28. How much heat must be absorbed to evaporate 16 g of NH_3 to its condensation point at −33 °C? (Use the heat of condensation for NH_3 = 1,380 J/g)

A. 86.5 J
B. 2,846 J
C. 118 J
D. 22,080 J
E. 1,380 J

29. Which of the reactions is the most exothermic, assuming that the following energy profiles have the same scale? (Use the notation of R = reactants and P = products)

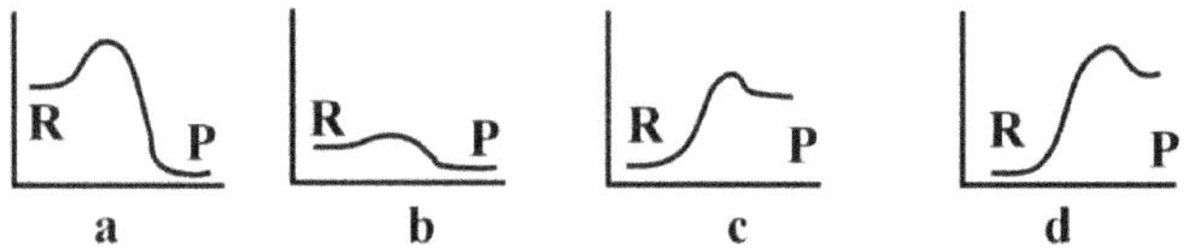

A. a
B. b
C. c
D. d
E. Cannot be determined

30. What is the heat of formation of NH_3 (*g*) of the following reaction:

$$2\ NH_3\ (g) \rightarrow N_2\ (g) + 3\ H_2\ (g)$$

(Use $\Delta H° = 92.4$ kJ/mol)

A. −92.4 kJ/mol
B. −184.4 kJ/mol
C. 46.2 kJ/mol
D. 92.4 kJ/mol
E. −46.2 kJ/mol

31. If a chemical reaction has a positive ΔH and a negative ΔS, the reaction tends to be:

A. at equilibrium
B. nonspontaneous
C. spontaneous
D. irreversible
E. unable to be determined

32. What happens to the entropy of a system as the components of the system are introduced to a larger number of possible arrangements, such as when liquid water transforms into water vapor?

A. Entropy of a system is solely dependent upon the amount of material undergoing reaction
B. Entropy of a system is independent of introducing the components of the system to a larger number of possible arrangements
C. Entropy increases because there are more ways for the energy to disperse
D. Entropy decreases because there are fewer ways in which the energy can disperse
E. Entropy increases because there are fewer ways for the energy to disperse

33. Which of the following quantities is needed to calculate the heat energy released as water turns to ice at 0 °C?

A. The heat of condensation for water and the mass
B. The heat of vaporization for water and the mass
C. The heat of fusion for water and the mass
D. The heat of solidification for water and the mass
E. The heat of fusion for water only

34. A nuclear power plant uses ^{235}U to convert water to steam that drives a turbine which turns a generator to produce electricity. What are the initial and final forms of energy, respectively?

A. Heat energy and electrical energy
B. Nuclear energy and electrical energy
C. Chemical energy and mechanical energy
D. Chemical energy and heat energy
E. Nuclear energy and mechanical energy

35. Which statement(s) is/are correct for the entropy?

I. Higher for a sample of gas than for the same sample of liquid
II. A measure of the disorder in a system
III. Available energy for conversion into mechanical work

A. I only
B. II only
C. III only
D. I and II only
E. I, II and III

36. Given the following data, what is the heat of formation for ethanol?

$C_2H_5OH + 3\ O_2 \rightarrow 2\ CO_2 + 3\ H_2O : \Delta H = 327.0$ kcal/mole

$H_2O \rightarrow H_2 + ½\ O_2 : \Delta H = +68.3$ kcal/mole

$C + O_2 \rightarrow CO_2 : \Delta H = -94.1$ kcal/mole

A. –720.1 kcal
B. –327.0 kcal
C. +62.6 kcal
D. +720.1 kcal
E. +327.0 kcal

37. Based on the reaction shown, which statement is true?

$S + O_2 \rightarrow SO_2 + 69.8$ kcal

A. 69.8 kcal are consumed when 32.1 g of sulfur reacts
B. 69.8 kcal are produced when 32.1 g of sulfur reacts
C. 69.8 kcal are consumed when 1 g of sulfur reacts
D. 69.8 kcal are produced when 1 g of sulfur reacts
E. 69.8 kcal are produced when 1 g of sulfur dioxide is produced

38. In the reaction, $2\ H_2\ (g) + O_2\ (g) \rightarrow 2\ H_2O\ (g)$, entropy is:

A. increasing
B. the same
C. inversely proportional
D. decreasing
E. unable to be determined

39. For an isolated system, which can NOT be exchanged between the system and its surroundings?

I. Temperature II. Matter III. Energy

A. I only
B. II only
C. III only
D. II and III only
E. I, II and III

40. Which is a state function?

I. ΔG II. ΔH III. ΔS

A. I only
B. II only
C. III only
D. I and II only
E. I, II and III

Notes for active learning

Practice Set 3: Questions 41–60

41. To simplify comparisons, the energy value of fuels is expressed in units of:

A. kcal/g
B. kcal/L
C. J/kcal
D. kcal/mol
E. kcal

42. Which process is slowed down when an office worker places a lid on a hot cup of coffee?

I. Radiation   II. Conduction   III. Convection

A. I only
B. II only
C. III only
D. I and III only
E. I, II and III

43. Which statement below is always true for a spontaneous chemical reaction?

A. $\Delta S_{sys} - \Delta S_{surr} = 0$
B. $\Delta S_{sys} + \Delta S_{surr} > 0$
C. $\Delta S_{sys} + \Delta S_{surr} < 0$
D. $\Delta S_{sys} + \Delta S_{surr} = 0$
E. $\Delta S_{sys} - \Delta S_{surr} < 0$

44. A fuel cell contains hydrogen and oxygen gas that react explosively, and the energy converts water to steam, which drives a turbine to turn a generator that produces electricity. What energy changes are employed in the process?

I. Mechanical → electrical energy
II. Heat → mechanical energy
III. Chemical → heat energy

A. I only
B. II only
C. I and II only
D. I and III only
E. I, II and III

45. Determine the value of ΔE°_{rxn} for this reaction, whereby the standard enthalpy of reaction (ΔH°_{rxn}) is –311.5 kJ mol^{-1}:

$C_2H_2\ (g) + 2\ H_2\ (g) \rightarrow C_2H_6\ (g)$

A. –306.5 kJ mol^{-1}
B. –318.0 kJ mol^{-1}
C. +346.0 kJ mol^{-1}
D. +306.5 kJ mol^{-1}
E. +466 kJ mol^{-1}

46. For a closed system, what can be exchanged between the system and its surroundings?

I. Heat II. Matter III. Energy

A. I only
B. II only
C. III only
D. I and III only
E. I, II and III

47. Which reaction is accompanied by an *increase* in entropy?

A. $Na_2CO_3\ (s) + CO_2\ (g) + H_2O\ (g) \rightarrow 2\ NaHCO_3\ (s)$
B. $BaO\ (s) + CO_2\ (g) \rightarrow BaCO_3\ (s)$
C. $CH_4\ (g) + H_2O\ (g) \rightarrow CO\ (g) + 3\ H_2\ (g)$
D. $ZnS\ (s) + 3/2\ O_2\ (g) \rightarrow ZnO\ (s) + SO_2\ (g)$
E. $N_2\ (g) + 3\ H_2\ (g) \rightarrow 2\ NH_3\ (g)$

48. Under standard conditions, which reaction has the largest difference between the energy of reaction and enthalpy?

A. C (*graphite*) → C (*diamond*)
B. C (*graphite*) + $O_2\ (g)$ → C (*diamond*)
C. 2 C (*graphite*) + $O_2\ (g) \rightarrow 2\ CO\ (g)$
D. C (*graphite*) + $O_2\ (g) \rightarrow CO_2\ (g)$
E. $CO\ (g) + NO_2\ (g) \rightarrow CO_2\ (g) + NO\ (g)$

49. Which type of reaction tends to be the most stable?

I. Isothermic II. Exergonic III. Endergonic

A. I only
B. II only
C. III only
D. I and II only
E. I and III only

50. Which is true for the thermodynamic functions G, H and S in $\Delta G = \Delta H - T\Delta S$?

A. G refers to the universe, H to the surroundings and S to the system
B. G, H, and S refer to the system
C. G and H refers to the surroundings and S to the system
D. G and H refer to the system and S to the surroundings
E. G and S refers to the system and H to the surroundings

51. Which of the following statements is true for the following reaction? (Use the change in enthalpy, $\Delta H° = -113.4$ kJ/mol and the change in entropy, $\Delta S° = -145.7$ J/K mol)

$2\ NO\ (g) + O_2\ (g) \rightarrow 2\ NO_2\ (g)$

A. The reaction is at equilibrium at 25 °C under standard conditions
B. The reaction is spontaneous at only high temperatures
C. The reaction is spontaneous only at low temperatures
D. The reaction is spontaneous at all temperatures
E. $\Delta G°$ becomes more favorable as temperature increases

52. Which of the following expressions defines enthalpy? (Use the conventions: q = heat, U = internal energy, P = pressure and V = volume)

A. $q - \Delta U$
B. $U + q$
C. q
D. ΔU
E. U + PV

53. Which statement is true regarding entropy?

I. It is a state function
II. It is an extensive property
III. It has an absolute zero value

A. I only
B. III only
C. I and II only
D. I and III only
E. I, II and III

54. Where does the energy released during an exothermic reaction originate from?

A. The kinetic energy of the surrounding
B. The kinetic energy of the reacting molecules
C. The potential energy of the reacting molecules
D. The thermal energy of the reactants
E. The potential energy of the surrounding

55. The species in the reaction $KClO_3$ (*s*) → KCl (*s*) + 3/2 O_2 (*g*) have the values for standard enthalpies of formation at 25 °C. At constant physical states, assume that the values of $\Delta H°$ and $\Delta S°$ are constant throughout a broad temperature range. Which of the following conditions may apply to the reaction? (Use $KClO_3$ (*s*) with $\Delta H_f° = -391.2$ kJ mol^{-1} and KCl (*s*) with $\Delta H_f° = -436.8$ kJ mol^{-1})

A. Nonspontaneous at low temperatures, but spontaneous at high temperatures
B. Spontaneous at low temperatures but nonspontaneous at high temperatures
C. Nonspontaneous at all temperatures over a broad temperature range
D. Spontaneous at all temperatures over a broad temperature range
E. No conclusion can be drawn about spontaneity based on the information

56. What is the standard enthalpy change for the reaction?

$$P_4\,(s) + 6\,Cl_2\,(g) \rightarrow 4\,PCl_3\,(l) \qquad \Delta H° = -1{,}289\text{ kJ}$$

$$3\,P_4\,(s) + 18\,Cl_2\,(g) \rightarrow 12\,PCl_3\,(l)$$

A. 426 kJ
B. −1,345 kJ
C. −366 kJ
D. 1,289 kJ
E. −3,867 kJ

57. Which of the following reactions is endothermic?

A. $PCl_3 + Cl_2 \rightarrow PCl_5 + \textit{heat}$
B. $2\,NO_2 \rightarrow N_2 + 2\,O_2 + \textit{heat}$
C. $CH_4 + NH_3 + \textit{heat} \rightarrow HCN + 3\,H_2$
D. $NH_3 + HBr \rightarrow NH_4Br$
E. $PCl_3 + Cl_2 \rightarrow PCl_5 + \textit{heat}$

58. When the system undergoes a spontaneous reaction, is it possible for the entropy of a system to decrease?

A. No, because this violates the second law of thermodynamics
B. No, because this violates the first law of thermodynamics
C. Yes, but only if the reaction is endothermic
D. Yes, but only if the entropy gain of the environment is greater than the entropy loss in the system
E. Yes, but only if the entropy gain of the environment is smaller than the entropy loss in the system

59. A 500 ml beaker of distilled water is placed under a bell jar, then covered by a layer of opaque insulation. After several days, some of the water evaporated. The contents of the bell jar are what kind of system?

A. endothermic
B. exergonic
C. closed
D. open
E. isolated

60. Which law explains the observation that the amount of heat transfer accompanying a change in one direction is equal in magnitude but opposite in sign to the amount of heat transfer in the opposite direction?

A. Law of Conservation of Mass
B. Law of Definite Proportions
C. Avogadro's Law
D. Boyle's Law
E. Law of Conservation of Energy

Notes for active learning

Practice Set 4: Questions 61–80

61. What is the term for a reaction that proceeds by releasing heat energy?

A. Endothermic reaction
B. Isothermal reaction
C. Exothermic reaction
D. Nonspontaneous
E. None of the above

62. If a stationary gas has a kinetic energy of 500 J at 25 °C, what is its kinetic energy at 50 °C?

A. 125 J
B. 450 J
C. 540 J
D. 1,120 J
E. 270 J

63. Which is NOT true for entropy in a closed system according to the equation $\Delta S = Q / T$?

A. Entropy is a measure of energy dispersal of the system
B. The equation is only valid for a reversible process
C. Changes due to heat transfer are greater at low temperatures
D. Disorder of the system decreases as heat is transferred out of the system
E. Increases as temperature decreases

64. In which of the following pairs of physical changes are both processes exothermic?

A. Melting and condensation
B. Freezing and condensation
C. Sublimation and evaporation
D. Freezing and sublimation
E. None of the above

65. A fuel cell contains hydrogen and oxygen gas that react explosively, and the energy converts water to steam which drives a turbine to turn a generator that produces electricity. What are the initial and final forms of energy, respectively?

A. Chemical and electrical energy
B. Nuclear and electrical energy
C. Chemical and mechanical energy
D. Chemical and heat energy
E. Nuclear and mechanical energy

66. Which must be true concerning a solution that reached an equilibrium where chemicals are mixed in a redox reaction?

A. $\Delta G° = \Delta G$
B. $E = 0$
C. $\Delta G° < 1$
D. $K = 1$
E. $K < 1$

67. A solid sample at room temperature spontaneously sublimes, forming a gas. This change in state is accompanied by which of the changes in the sample?

A. Entropy decreases and energy increases
B. Entropy increases and energy decreases
C. Entropy and energy decrease
D. Entropy and energy increase
E. Entropy and energy are equal

68. At constant temperature and pressure, a negative ΔG indicates that the:

A. reaction is nonspontaneous
B. reaction is fast
C. reaction is spontaneous
D. reaction is endothermic
E. $\Delta S > 0$

69. The process of H_2O (*g*) → H_2O (*l*) is nonspontaneous under pressure of 760 torr and temperatures of 378 K because:

A. $\Delta H = T\Delta S$
B. $\Delta G < 0$
C. $\Delta H > 0$
D. $\Delta H < T\Delta S$
E. $\Delta H > T\Delta S$

70. Consider the contribution of entropy to the spontaneity of the reaction:

$$2\ Al_2O_3\ (s) \rightarrow 4\ Al\ (s) + 3\ O_2\ (g),\ \Delta G = +138\ \text{kcal}.$$

As written, the reaction is [ ] and the entropy of the system [ ].

A. non-spontaneous… decreases
B. non-spontaneous… increases
C. spontaneous… decreases
D. spontaneous… increases
E. non-spontaneous… does not change

71. The ΔG of a reaction is the maximum energy that the reaction releases to do:

A. P–V work only
B. work and release heat
C. any type of work
D. non P–V work only
E. work and generate heat

72. Which of the following represent forms of internal energy?

I. bond energy
II. thermal energy
III. gravitational energy

A. I only
B. II only
C. I and II only
D. I and III only
E. I, II and III

73. If it takes energy to break bonds and energy is gained in forming bonds, how can some reactions be exothermic while others are endothermic?

A. Some products have more energy than others and require energy to be formed
B. Some reactants have more energetic bonds than others and release energy
C. It is the number of bonds that is determinative. Since bonds have the same amount of energy, the net gain or net loss of energy depends on the number of bonds
D. It is the amount of energy that is determinative. Some bonds are stronger than others, so there is a net gain or net loss of energy when formed
E. None of the above

74. Which constant is represented by A in the following calculation to determine how much heat is required to convert 60 g of ice at −25 °C to steam at 320 °C?

$$\text{Total heat} = [(A)\cdot(60\text{ g})\cdot(25\text{ °C})] + [(\text{heat of fusion})\cdot(60\text{ g})] +$$
$$+ [(4.18\text{ J/g}\cdot\text{°C})\cdot(60\text{ g})\cdot(100\text{ °C})] + [(B)\cdot(60\text{ g})] + [(C)\cdot(60\text{ g})\cdot(220\text{ °C})]$$

A. Specific heat of water
B. The heat of vaporization of water
C. The heat of condensation
D. The heat capacity of steam
E. Specific heat of ice

75. The heat of formation of water vapor is:

A. positive, but greater than the heat of formation for H_2O (*l*)
B. positive and smaller than the heat of formation for H_2O (*l*)
C. negative, but greater than the heat of formation for H_2O (*l*)
D. negative and smaller than the heat of formation for H_2O (*l*)
E. positive and equal to the heat of formation for H_2O (*l*)

76. Entropy can be defined as the amount of:

A. equilibrium in a system
B. chemical bonds that are changed during a reaction
C. energy required to initiate a reaction
D. energy required to rearrange chemical bonds
E. disorder in a system

77. Which is true of an atomic fission bomb according to the conservation of mass and energy law?

A. The mass of the bomb and the fission products are identical
B. A small amount of mass is converted into energy
C. The energy of the bomb and the fission products are identical
D. The mass of the fission bomb is greater than the mass of the products
E. None of the above

78. Once an object enters a black hole, astronomers consider it to have left the universe which means the universe is:

A. entropic
B. isolated
C. closed
D. open
E. exergonic

79. Which ranking from lowest to highest entropy per gram of NaCl is correct?

A. NaCl (*s*) < NaCl (*l*) < NaCl (*aq*) < NaCl (*g*)
B. NaCl (*s*) < NaCl (*l*) < NaCl (*g*) < NaCl (*aq*)
C. NaCl (*g*) < NaCl (*aq*) < NaCl (*l*) < NaCl (*s*)
D. NaCl (*s*), NaCl (*aq*), NaCl (*l*), NaCl (*g*)
E. NaCl (*l*) < NaCl (*aq*) < NaCl (*g*) < NaCl (*s*)

80. From the given bond energies, how many kJ of energy are released or absorbed from the reaction of one mole of N_2 with three moles of H_2 to form two moles of NH_3?

$$N{\equiv}N + H{-}H + H{-}H + H{-}H \rightarrow NH_3 + NH_3$$

H–N: 389 kJ/mol H–H: 436 kJ/mol N≡N: 946 kJ/mol

A. –80 kJ/mol released
B. +89.5 kJ/mol absorbed
C. –946 kJ/mol released
D. +895 kJ/mol absorbed

Notes or active learning

Notes or active learning

Answer Key & Detailed Explanations

Answer Keys

1: A	11: C	21: A	31: B	41: A	51: C	61: C	71: D
2: B	12: A	22: C	32: C	42: C	52: E	62: C	72: C
3: E	13: B	23: A	33: D	43: B	53: C	63: C	73: D
4: E	14: D	24: B	34: B	44: E	54: C	64: B	74: E
5: A	15: C	25: C	35: D	45: A	55: D	65: A	75: C
6: D	16: B	26: C	36: A	46: D	56: E	66: B	76: E
7: E	17: A	27: A	37: B	47: C	57: C	67: D	77: B
8: A	18: B	28: D	38: D	48: C	58: D	68: C	78: D
9: C	19: C	29: A	39: E	49: B	59: E	69: E	79: A
10: E	20: C	30: E	40: E	50: B	60: E	70: B	80: A

===

Practice Set 1: Questions 1–20

===

1. A is correct.

A reaction's enthalpy is specified by ΔH (not ΔG). ΔH determines whether a reaction is exothermic (releases heat to surroundings) or endothermic (absorbs heat from surroundings).

To predict spontaneity of reaction, use the Gibbs free energy equation:

$$\Delta G = \Delta H^\circ - T\Delta S$$

The reaction is spontaneous (i.e., exergonic) if ΔG is negative.

The reaction is nonspontaneous (i.e., endergonic) if ΔG is positive.

2. B is correct.

Heat is energy, and the energy from the heat is transferred to the gas molecules, which increases their movement (i.e., kinetic energy).

Temperature is a measure of the average kinetic energy of the molecules.

3. E is correct.

Heat capacity is the amount of heat required to increase the temperature of *the whole sample* by 1 °C.

Specific heat is the heat required to increase the temperature of *1 gram* of sample by 1 °C.

Heat = mass × specific heat × change in temperature:

$$q = m \times c \times \Delta T$$

$$q = 21.0\ g \times 0.382\ J/g{\cdot}°C \times (68.5\ °C - 21.0\ °C)$$

$$q = 21.0\ g \times 0.382\ J/g{\cdot}°C \times (47.5\ °C)$$

$$q = 381\ J$$

4. E is correct.

This is a theory/memorization question.

However, the problem can be solved by comparing the options for the most plausible.

Kinetic energy is correlated with temperature, so the formula that involves temperature would be a good choice.

Another approach is to analyze the units. Apply the units to the variables and evaluate them to obtain the answer.

For example, choices A and B have moles, pressure (Pa, bar or atm), and area (m^2), so it is not possible for them to result in energy (J) when multiplied.

Options C and D have molarity (moles/L) and volume (L or m^3); they cannot result in energy (J) when multiplied.

Option E is nRT: mol × (J/mol K) × K; units but J cancel, which is a plausible formula for kinetic energy.

Additionally, dimensional analysis limits the answer choices on this type of question for questions using formulas.

5. A is correct.

ΔH refers to enthalpy (or heat).

Endothermic reactions have heat as a reactant.

Exothermic reactions have heat as a product.

Exothermic reactions release heat and cause the temperature of the immediate surroundings to rise (i.e., a net loss of energy).

Endothermic reactions absorb heat and cool the surroundings (i.e., a net gain of energy).

Endothermic reactions absorb energy to break strong bonds to form a less stable state (i.e., positive enthalpy).
Exothermic reactions release energy when forming stronger bonds to produce a more stable state (i.e., negative enthalpy).

6. D is correct.

Endothermic reactions absorb energy to break strong bonds to form a less stable state (i.e., positive enthalpy).
Exothermic reactions release energy when forming stronger bonds to produce a more stable state (i.e., negative enthalpy).

The spontaneity of a reaction is determined by the calculation of ΔG using:

$$\Delta G = \Delta H - \mathrm{T}\Delta S$$

The reaction is spontaneous when the value of ΔG is negative.

In this problem, the reaction is endothermic, which means that ΔH is positive.

The reaction decreases S, which means that ΔS value is negative.

Substituting these values into the equation:

$$\Delta G = \Delta H - \mathrm{T}(-\Delta S)$$

$\Delta G = \Delta H + \mathrm{T}\Delta S$, with ΔH and ΔS both positive values

For this problem, the value of ΔG is positive.

The reaction does not occur because ΔG needs to be negative for a spontaneous reaction to proceed, the products are more stable than the reactants.

7. E is correct.

Chemical energy is potential energy stored in molecular bonds.

Electrical energy is the kinetic energy associated with the motion of electrons (e.g., in wires, circuits, lightning). The potential energy stored in a flashlight battery is electrical energy.

Heat is the spontaneous transfer of energy from a hot object to a cold one with no displacement or deformation of the objects (i.e., no work done).

8. A is correct.

Entropy is higher for less organized states (i.e., more random or disordered).

9. C is correct.

ΔH refers to enthalpy (or heat).

Endothermic reactions have heat as a reactant. Exothermic reactions have heat as a product.

Exothermic reactions release heat and cause the temperature of the immediate surroundings to rise (i.e., a net loss of energy). An endothermic process absorbs heat and cools the surroundings (i.e., a net energy gain).

Endothermic reactions absorb energy to break strong bonds to form a less stable state (i.e., positive enthalpy).

Exothermic reactions release energy when forming stronger bonds to produce a more stable state (i.e., negative enthalpy).

The reaction is nonspontaneous (i.e., endergonic) if the products are less stable than the reactants and ΔG is positive.

The reaction is spontaneous (i.e., exergonic) if the products are more stable than the reactants and ΔG is negative.

10. E is correct.

This cannot be determined when the given ΔG = ΔH – TΔS, terms cancel.

The system is at equilibrium when ΔG = 0, but in this question, both sides of the equation cancel:

$$\underbrace{X - RY}_{\Delta G\text{ term}} = \underbrace{X - RY}_{\substack{\Delta H - T\Delta S\\ \text{term}}}, \quad \text{so } 0 = 0$$

11. C is correct.

Gibbs free energy: $\Delta G = \Delta H - T\Delta S$

Stable molecules are spontaneous because they have a negative (or relatively low) ΔG.

ΔG is most negative (most stable) when ΔH is the smallest and ΔS is largest.

12. A is correct.

ΔH refers to enthalpy (or heat).

Endothermic reactions have heat as a reactant.

Exothermic reactions have heat as a product.

The reaction is nonspontaneous (i.e., endergonic) if the products are less stable than the reactants and ΔG is positive.

The reaction is spontaneous (i.e., exergonic) if the products are more stable than the reactants and ΔG is negative.

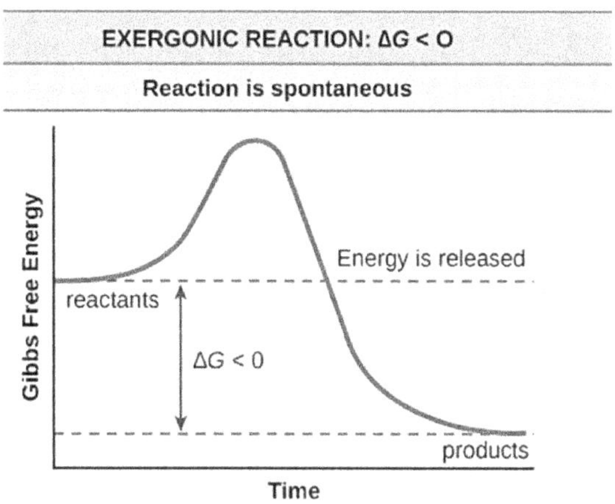

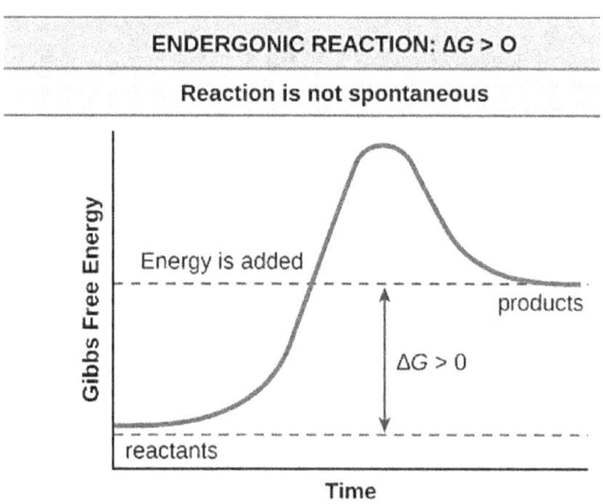

continued...

Exothermic reactions release heat and cause the temperature of the immediate surroundings to rise (i.e., a net loss of energy).

Endothermic reactions absorb heat and cool the surroundings (i.e., a net gain of energy).

Endothermic reactions absorb energy to break strong bonds to form a less stable state (i.e., positive enthalpy).

Exothermic reactions release energy when forming stronger bonds to produce a more stable state (i.e., negative enthalpy).

13. B is correct.

Gibbs free energy:

$$\Delta G = \Delta H - T\Delta S$$

Entropy (ΔS) determines the favorability of chemical reactions.

If entropy is large, the reaction is more likely to proceed.

14. D is correct.

Bonds within reactants are broken in a chemical reaction; new bonds form to create products.

Therefore, in bond dissociation problems,

ΔH reaction = sum of bond energy in reactants – the sum of bond energy in products

For $H_2C{=}CH_2 + H_2 \rightarrow CH_3{-}CH_3$:

$\Delta H_{reaction}$ = sum of bond energy in reactants – sum of bond energy in products

$$\Delta H_{reaction} = [(C{=}C) + 4(C{-}H) + (H{-}H)] - [(C{-}C) + 6(C{-}H)]$$

$$\Delta H_{reaction} = [612\text{ kJ} + (4 \times 412\text{ kJ}) + 436\text{ kJ}] - [348\text{ kJ} + (6 \times 412\text{ kJ})]$$

$$\Delta H_{reaction} = -124\text{ kJ}$$

Remember that this is the opposite of ΔH_f problems, where:

$$\Delta H_{reaction} = (\text{sum of } \Delta H_f \text{ products}) - (\text{sum of } \Delta H_f \text{ reactants})$$

15. C is correct.

Bond dissociation energy is the energy required to break a bond between two gaseous atoms and helps estimate the enthalpy change in a reaction.

16. B is correct.

The ΔH value is positive, which means the reaction is endothermic and absorbs energy from the surroundings.

ΔH value is expressed as the energy released/absorbed per mole (for species with a coefficient of 1).

If the coefficient is 2, it is the energy transferred per 2 moles.

For example, 2 moles of NO are produced from 1 mole for each reagent for the stated reaction.

Therefore, 43.2 kcal is produced when 2 moles of NO are produced.

17. A is correct.

State functions depend only on the initial and final states of the system and are independent of the paths taken to reach the final state.

Common thermodynamics state functions include the internal energy, enthalpy, entropy, pressure, temperature, and volume.

Work and heat relate to the change in energy of a system when it moves from one state to another, depending on how a system changes between states.

18. B is correct.

The temperature (i.e., average kinetic energy) of a substance remains constant during a phase change (e.g., solid to liquid or liquid to gas).

For example, the heat (i.e., energy) breaks bonds between the ice molecules and changes phases into the liquid phase.

Since the average kinetic energy of the molecules does not change at the moment of the phase change (i.e., melting), the temperature of the molecules does not change.

19. C is correct.

Bonds within reactants are broken in a chemical reaction, and new bonds form to create products.

Bond dissociation:

$$\Delta H_{reaction} = \text{(sum of bond energy in reactants)} - \text{(sum of bond energy in products)}$$

$$O{=}C{=}O + 3\ H_2 \rightarrow CH_3{-}O{-}H + H{-}O{-}H$$

$$\Delta H_{reaction} = \text{(sum of bond energy in reactants)} - \text{(sum of bond energy in products)}$$

$$\Delta H_{reaction} = [2(C{=}O) + 3(H{-}H)] - [3(C{-}H) + (C{-}O) + (O{-}H) + 2(O{-}H)]$$

$$\Delta H_{reaction} = [(2 \times 743\ kJ) + (3 \times 436\ kJ)] - [(3 \times 412\ kJ) + 360\ kJ + 463\ kJ + (2 \times 463\ kJ)]$$

$$\Delta H_{reaction} = -191\ kJ$$

This is the reverse of ΔH_f problems, where:

$$\Delta H_{reaction} = (\text{sum of } \Delta H_{f\ product}) - (\text{sum of } \Delta H_{f\ reactant})$$

20. C is correct.

Entropy indicates how a system is organized; increased entropy means less order.

The entropy of the universe is continuously increasing because a system moves towards more disorder unless energy is added to the system.

The second law of thermodynamics states that the entropy of interconnected systems, without the addition of energy, always increases.

===

Practice Set 2: Questions 21–40

===

21. A is correct.

Gibbs free energy: $\Delta G = \Delta H - T\Delta S$

The value that is the largest positive value is the most endothermic, while the value that is the largest negative value is the most exothermic.

ΔH refers to enthalpy (or heat).

Endothermic reactions have heat as a *reactant.*

Exothermic reactions have heat as a *product.*

Exothermic reactions release heat and cause the temperature of the immediate surroundings to rise (i.e., a net loss of energy). An endothermic process absorbs heat and cools the surroundings (i.e., a net gain of energy).

Endothermic reactions absorb energy to break strong bonds to form a less stable state (i.e., positive enthalpy).

Exothermic reactions release energy when forming stronger bonds to produce a more stable state (i.e., negative enthalpy).

22. C is correct.

Potential Energy **Stored energy and the energy of position (gravitational)**	**Kinetic Energy** **Energy of motion: motion of waves, electrons, atoms, molecules, and substances**
Chemical Energy Chemical energy is the energy stored in the bonds of atoms and molecules. Examples of stored chemical energy: biomass, petroleum, natural gas, propane, coal.	**Thermal Energy** Thermal energy is internal energy in substances; it is the vibration and movement of atoms and molecules. Geothermal energy is an example of thermal energy.

23. A is correct.

By convention, the ΔG for an element in its standard state is 0.

24. B is correct.

Test strategy: when given a choice among similar explanations, carefully evaluate the most descriptive one.

However, check the statement because sometimes the most descriptive option is inaccurate.

Here, the longest option is accurate, which would be the best choice among all options.

25. C is correct.

ΔG is negative for spontaneous reactions.

26. C is correct.

An insulator reduces conduction (i.e., thermal energy transfer through matter).

Air and vacuum are excellent insulators. Storm windows, which have air wedged between two glass panes, work by utilizing this conduction principle.

27. A is correct.

Δ*H* refers to enthalpy (or heat).

Endothermic reactions have heat as a reactant. Exothermic reactions have heat as a product.

Exothermic reactions release heat and cause the temperature of the immediate surroundings to rise (i.e., a net loss of energy).

Endothermic reactions absorb heat and cool the surroundings (i.e., a net gain of energy).

The reaction is nonspontaneous (i.e., endergonic) if the products are less stable than the reactants and Δ*G* is positive. The reaction is spontaneous (i.e., exergonic) if the products are more stable than the reactants and Δ*G* is negative.

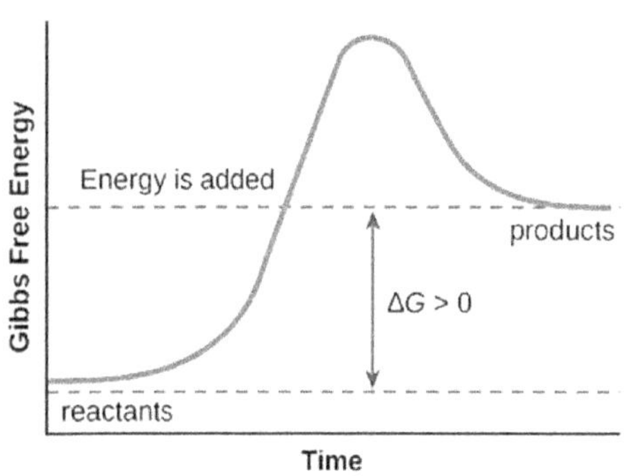

Endergonic reactions are nonspontaneous, and have products with more energy than the reactants

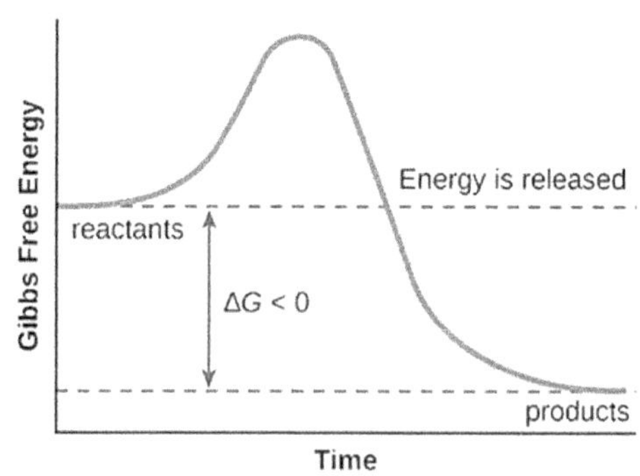

Exergonic reactions are spontaneous and have reactants with more energy than the products

28. D is correct.

Heat capacity is the heat required to increase the temperature of *the whole sample* by 1 °C.

Specific heat is the heat required to increase the temperature of *1 gram* of sample by 1 °C.

q = mass × heat of condensation

q = 16 g × 1,380 J/g

q = 22,080 J

29. A is correct.

EXERGONIC REACTION: ΔG < O

Reaction is spontaneous

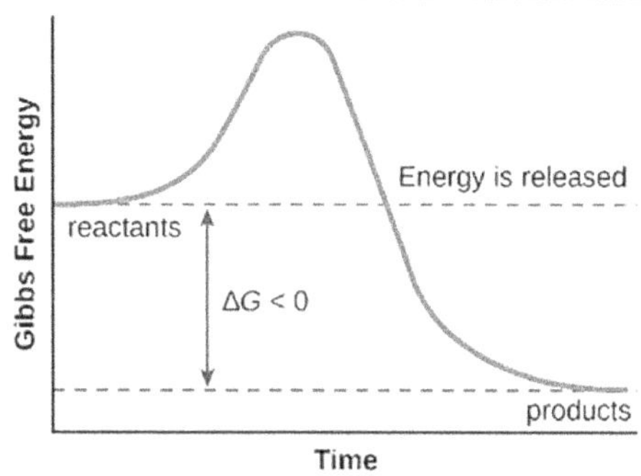

ENDERGONIC REACTION: ΔG > O

Reaction is not spontaneous

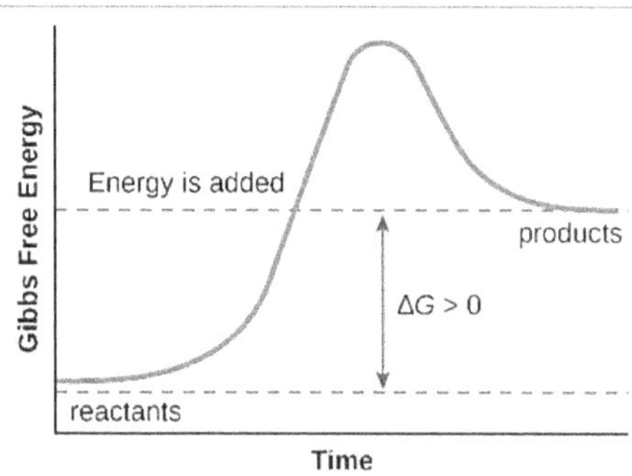

$\Delta G = \Delta H - T\Delta S$ refers to exergonic when ΔG is negative and endergonic when ΔG is positive.

If ΔS is 0, then $\Delta G = \Delta H$ because the $T\Delta S$ term cancels because it equals zero).

ΔH refers to exothermic when ΔH is negative and endothermic when ΔH is positive.

30. E is correct.

Heat of formation is the heat required or released upon creating one mole of the substance from its elements.

If the given reaction is reversed and divided by two, it would be the formation reaction of NH_3:

$$\frac{1}{2} N_2 + 3/2\ H_2 \rightarrow NH_3$$

ΔH of this reaction would be reversed (i.e., plus to minus sign) and divided by two:

$$\Delta H = -(92.4\ \text{kJ/mol}) / 2$$

$$\Delta H = -46.2\ \text{kJ/mol}$$

31. B is correct.

Gibbs free energy:

$$\Delta G = \Delta H - T\Delta S$$

With a positive ΔH and negative ΔS, ΔG is positive, so the reaction is nonspontaneous.

32. C is correct.

Entropy (ΔS) measures the degree of disorder in a system.

When a change causes the components of a system to transform into a higher energy phase (solid → liquid, liquid → gas), the entropy of the system increases.

33. D is correct.

Only the mass and heat of specific phase changes are required to calculate the energy released or absorbed during phase changes.

Because water is turning into ice, the specific heat required is the heat of solidification.

34. B is correct.

Nuclear energy is considered energy because splitting or combining atoms results in enormous amounts of energy used in atomic bombs and nuclear power plants.

Potential Energy **Stored energy and the energy of position (gravitational)**	**Kinetic Energy** **Energy of motion: motion of waves, electrons, atoms, molecules, and substances**
Nuclear Energy Nuclear energy is the energy stored in the nucleus of an atom. It is the energy that holds the nucleus together. The nucleus of the uranium atom is an example of nuclear energy.	**Electrical Energy** Electrical energy is the movement of electrons. Lightning and electricity are examples of electrical energy.

35. D is correct.

Entropy is a measurement of disorder.

Gases contain more energy than liquids of the same element and have a higher degree of randomness.

Entropy is the unavailable energy that cannot be converted into mechanical work.

36. A is correct.

There are two different methods available to solve this problem.

Both methods are based on the definition of ΔH formation as energy consumed/released when one mole of the molecule is produced from its elemental atoms.

Method 1

Rearrange and add the equations to create the formation reaction of C_2H_5OH.

Information provided by the problem:

$$C_2H_5OH + 3\ O_2 \rightarrow 2\ CO_2 + 3\ H_2O \qquad \Delta H = 327\ \text{kcal}$$

$$H_2O \rightarrow H_2 + \tfrac{1}{2}\ O_2 \qquad \Delta H = 68.3\ \text{kcal}$$

$$C + O_2 \rightarrow CO_2 \qquad \Delta H = -94.1\ \text{kcal}$$

Place C_2H_5OH on the product side. For the other two reactions, arrange them so the elements are on the left and the molecules are on the right (remember that the goal is to create a formation reaction: elements forming a molecule).

When reversing the direction of a reaction, change the positive/negative sign of ΔH.

Multiplying the whole reaction by a coefficient would multiply ΔH by the same ratio.

$2\ CO_2 + 3\ H_2O \rightarrow C_2H_5OH + 3\ O_2$ $\quad \Delta H = -327$ kcal

$3\ H_2 + 3/2\ O_2 \rightarrow 3\ H_2O$ $\quad \Delta H = -204.9$ kcal

$\underline{2\ C + 2\ O_2 \rightarrow 2\ CO_2 \quad \Delta H = -188.2 \text{ kcal}}$

$2\ C + 3\ H_2 + 1/2\ O_2 \rightarrow C_2H_5OH$ $\quad \Delta H = -720.1$ kcal

ΔH formation of C_2H_5OH is –720.1 kcal.

Method 2

The heat of formation (ΔH_f) data can be used to calculate ΔH of a reaction:

$$\Delta H_{reaction} = \text{sum of } \Delta H_{f\ product} - \text{sum of } \Delta H_{f\ reactant}$$

Apply the formula on the first equation:

$$C_2H_5OH + 3\ O_2 \rightarrow 2\ CO_2 + 3\ H_2O \quad \Delta H = 327 \text{ kcal}$$

$$\Delta H_{reaction} = \text{sum of } \Delta H_{f\ product} - \text{sum of } \Delta H_{f\ reactant}$$

$$327 \text{ kcal} = [2(\Delta H_f\ CO_2) + 3(\Delta H_f\ H_2O)] - [(\Delta H_f\ C_2H_5OH) + 3(\Delta H_f\ O_2)]$$

Oxygen (O_2) is an element, so $\Delta H_f = 0$

For CO_2 and H_2O, ΔH_f information can be obtained from the other 2 reactions:

$H_2O \rightarrow H_2 + \frac{1}{2}\ O_2$ $\quad \Delta H = 68.3$ kcal

$C + O_2 \rightarrow CO_2$ $\quad \Delta H = -94.1$ kcal

The CO_2 reaction is already a formation reaction – creating one mole of a molecule from its elements.

Therefore, ΔH_f of CO_2 = –94.1 kcal

If the H_2O reaction is reversed, it will be a formation reaction.

$H_2 + \frac{1}{2}\ O_2 \rightarrow H_2O$ $\quad \Delta H = -68.3$ kcal

Using those values, calculate ΔH_f of C_2H_5OH:

$$327 \text{ kcal} = [2(\Delta H_f\ CO_2) + 3(\Delta H_f\ H_2O)] - [(\Delta H_f\ C_2H_5OH) + 3(\Delta H_f\ O_2)]$$

$$327 \text{ kcal} = [2(-94.1 \text{ kcal}) + 3(-68.3 \text{ kcal})] - [(\Delta H_f\ C_2H_5OH) + 3(0 \text{ kcal})]$$

$$327 \text{ kcal} = [(-188.2 \text{ kcal}) + (-204.9 \text{ kcal})] - (\Delta H_f\ C_2H_5OH)$$

$$327 \text{ kcal} = (-393.1 \text{ kcal}) - (\Delta H_f\ C_2H_5OH)$$

$$\Delta H_f\ C_2H_5OH = -720.1 \text{ kcal}$$

37. B is correct.

$$S + O_2 \rightarrow SO_2 + 69.8 \text{ kcal}$$

The reaction indicates that 69.8 kcal of energy is released (i.e., on the product side).

All species in this reaction have the coefficient of 1, which means that ΔH is calculated by reacting 1 mole of each reagent to create 1 mole of product.

The atomic mass of sulfur is 32.1 g, which means that this reaction uses 32.1 g of sulfur.

38. D is correct.

The disorder of the system is decreasing as more complex and ordered molecules are forming from the reaction of H_2 and O_2 gases.

3 moles of gas combine to form 2 moles of gas.

Additionally, two molecules are joining to form one molecule.

39. E is correct.

In thermodynamics, an isolated system is enclosed by rigid, immovable walls through which neither matter nor energy passes.

Temperature is a measure of energy and is not a form of energy. Therefore, the temperature cannot be exchanged between the system and the surroundings.

A closed system can exchange energy (as heat or work) with its surroundings, but not matter.

An isolated system cannot exchange energy (as heat or work) or matter with the surroundings.

An open system can exchange energy and matter with the surroundings.

40. E is correct.

State functions depend only on the initial and final states of the system and are independent of the paths taken to reach the final state.

Common examples of a state function in thermodynamics include internal energy, enthalpy, entropy, pressure, temperature, and volume.

Notes for active learning

==

Practice Set 3: Questions 41–60

==

41. A is correct.

When comparing various fuels in varying forms (e.g., solid, liquid, gas), it is easiest to compare energy released in terms of the mass because matter has mass regardless of its state (as opposed to volume, which is convenient for a liquid or gas, but not for a solid).

Moles are more complicated because they have to be converted to mass before a direct comparison can be made between the fuels.

42. C is correct.

Convection is heat carried by fluids (i.e., liquids and gases).

Heat is prevented from leaving the system through convection when the lid is placed on the cup.

43. B is correct.

For a reaction to be spontaneous, the entropy ($\Delta S_{system} + \Delta S_{surrounding}$) has to be positive (i.e., greater than 0).

44. E is correct.

Explosion of gases:

chemical → heat energy

Heat converting into steam, which in turn moves a turbine:

heat → mechanical energy

Generator creates electricity:

mechanical → electrical energy

Potential Energy **Stored energy and the energy of position (gravitational)**	**Kinetic Energy** **Energy of motion: motion of waves, electrons, atoms, molecules, and substances**
Chemical Energy Chemical energy is the energy stored in the bonds of atoms and molecules. Examples include stored chemical energy: biomass, petroleum, natural gas, propane, coal. **Nuclear Energy** Nuclear energy is the energy stored in the nucleus of an atom. It is the energy that holds the nucleus. The nucleus of the uranium atom is an example of nuclear energy. **Stored Mechanical Energy** Stored mechanical energy is energy stored in objects by applying force. Compressed springs and stretched rubber bands are examples of stored mechanical energy. **Gravitational Energy** Gravitational Energy is the energy of place or position. Water in a reservoir behind a hydropower dam is an example of gravitational potential energy. When the water is released to spin the turbines, it becomes kinetic energy.	**Radiant Energy** Radiant energy is electromagnetic energy that travels in transverse waves. Radiant energy includes visible light, x-rays, gamma rays, and radio waves. Solar energy is an example of radiant energy. **Thermal Energy** Thermal energy is internal energy in substances; it is the vibration and movement of atoms and molecules. Geothermal energy is an example of thermal energy. **Motion** The movement of objects or substances from one place to another is motion. Wind and hydropower are examples of motion. **Sound** Sound is the movement of energy through substances in longitudinal (compression/rarefaction) waves. **Electrical Energy** Electrical energy is the movement of electrons. Lightning and electricity are examples of electrical energy.

45. A is correct.

The relationship between enthalpy (ΔH) and internal energy (ΔE):

Enthalpy = Internal energy + work (for gases, work = PV)

$\Delta H = \Delta E + \Delta(PV)$

Solving for ΔE:

$\Delta E = \Delta H - \Delta(PV)$

According to ideal gas law:

PV = nRT

Substitute ideal gas law to the previous equation:

$\Delta E = \Delta H - \Delta(nRT)$

R and T are constant, which leaves Δn as the variable.

The reaction is $C_2H_2\ (g) + 2H_2\ (g) \rightarrow C_2H_6\ (g)$.

There are three gas molecules on the left and one on the right:

$\Delta n = 1 - 3 = -2$

In this problem, the temperature is not provided. However, the presence of degree symbols ($\Delta G°$, $\Delta H°$, $\Delta S°$) indicates that those are standard values, measured at 25 °C or 298.15 K.

Double-check the units before performing calculations; ΔH is in kilojoules, while the gas constant (R) is 8.314 J/mol K. Convert ΔH to joules before calculating.

Use the Δn and T values to calculate ΔE:

$\Delta E = \Delta H - \Delta nRT$

$\Delta E = -311{,}500\ \text{J/mol} - (-2 \times 8.314\ \text{J/mol K} \times 298.15\ \text{K})$

$\Delta E = -306{,}542\ \text{J} \approx -306.5\ \text{kJ}$

46. D is correct.

A closed system can exchange energy (as heat or work) but not matter with its surroundings.

An isolated system cannot exchange energy (as heat or work) or matter with the surroundings.

An open system can exchange energy and matter with the surroundings.

47. C is correct.

An increase in entropy indicates an increase in disorder; find a reaction that creates a higher number of molecules on the product side than the reactant side.

48. C is correct.

When enthalpy is calculated, work done by gases is not taken into account; however, the energy of the reaction includes the work done by gases.

The largest difference between the energy of the reaction and enthalpy would be for reactions with the largest difference in the number of gas molecules between the product and reactants.

49. B is correct.

ΔH refers to enthalpy (or heat).

Endothermic reactions have heat as a *reactant.*

Exothermic reactions have heat as a *product.*

Endothermic reactions *absorb energy* to break strong bonds to form a less stable state (i.e., positive enthalpy).

Exothermic reactions *release energy* when forming stronger bonds to produce a more stable state (i.e., negative enthalpy).

The reaction is *nonspontaneous* (i.e., endergonic) if the products are less stable than reactants and ΔG is positive.

The reaction is *spontaneous* (i.e., exergonic) if the products are more stable than the reactants and ΔG is negative.

50. B is correct.

All thermodynamic functions in $\Delta G = \Delta H - T\Delta S$ refer to the system.

51. C is correct.

To predict spontaneity of reaction, use the Gibbs free energy equation:

$$\Delta G = \Delta H° - T\Delta S$$

The reaction is spontaneous if ΔG is negative.

Substitute the given values to the equation:

$$\Delta G = -113.4 \text{ kJ/mol} - [T \times (-145.7 \text{ J/K mol})]$$

$$\Delta G = -113.4 \text{ kJ/mol} + (T \times 145.7 \text{ J/K mol})$$

Important: note that the units are not identical, $\Delta H°$ is in kJ and $\Delta S°$ is in J.

Convert kJ to J (1 kJ = 1,000 J):

$$\Delta G = -113{,}400 \text{ J/mol} + (T \times 145.7 \text{ J/K mol})$$

It can be predicted that the value of ΔG would be negative if the value of T is small.

If T goes higher, ΔG approaches a positive value, and the reaction would be nonspontaneous.

52. E is correct.

Enthalpy: U + PV

53. C is correct.

State functions depend only on the initial and final states of the system and are independent of the paths taken to reach the final state.

Common thermodynamics state functions include: internal energy, enthalpy, entropy, pressure, temperature, and volume.

An extensive property is a property that changes when the size of the sample changes. Examples include mass, volume, length, and charge.

Entropy has no absolute zero value, and a substance at zero Kelvin has zero entropy (i.e., no motion).

54. C is correct.

In an exothermic reaction, the bonds formed are stronger than the bonds broken.

55. D is correct.

Calculate the value of $\Delta H_f = \Delta H_{f\,product} - \Delta H_{f\,reactant}$

$$\Delta H_f = -436.8\ \text{kJ mol}^{-1} - (-391.2\ \text{kJ mol}^{-1})$$

$$\Delta H_f = -45.6\ \text{kJ mol}^{-1}$$

To determine spontaneity, calculate the Gibbs free energy:

$$\Delta G = \Delta H° - T\Delta S$$

The reaction is spontaneous if ΔG is negative:

$$\Delta G = -45.6\ \text{kJ} - T\Delta S$$

Typical ΔS values are around 100–200 J.

If the ΔH value is –45.6 kJ or –45,600 J, the value of ΔG would still be negative unless $T\Delta S$ is less than –45,600.

Therefore, the reaction would be spontaneous over a broad range of temperatures.

56. E is correct.

Enthalpy is an extensive property: it varies with the quantity of the matter.

Reaction 1 : $P_4 + 6\ Cl_2 \rightarrow 4\ PCl_3$

$$\Delta H = -1{,}289\ \text{kJ}$$

Reaction 2 : $3\ P_4 + 18\ Cl_2 \rightarrow 12\ PCl_3$

$$\Delta H = ?$$

The only difference between reactions 1 and 2 is the coefficients.

In reaction 2, coefficients are three times reaction 1. Reaction 2 consumes three times as many reactants as reaction 1; reaction 2 releases three times as much energy as reaction 1.

$$\Delta H = 3 \times -1{,}289\ \text{kJ}$$

$$\Delta H = -3{,}867\ \text{kJ}$$

57. C is correct.

ΔH refers to enthalpy (or heat).

Endothermic reactions have heat as a reactant.

Exothermic reactions have heat as a product.

Endothermic reactions absorb energy to break strong bonds to form a less stable state (i.e., positive enthalpy).

Since endothermic reactions consume energy, there should be heat on the reactants' side.

Exothermic reactions *release energy* when forming stronger bonds to produce a more stable state (i.e., negative enthalpy).

The reaction is *nonspontaneous* (i.e., endergonic) if the *products are less stable* than the reactants and ΔG is positive.

The reaction is *spontaneous* (i.e., exergonic) if the *products are more stable* than the reactants and ΔG is negative.

58. D is correct.

According to the second law of thermodynamics, the entropy gain of the universe must be positive (i.e., increased disorder).

The entropy of a system, however, can be negative if the surroundings experience an increase in entropy greater than the negative entropy change experienced by the system.

59. E is correct.

In thermodynamics, an isolated system is enclosed by rigid, immovable walls through which neither matter nor energy passes.

Bell jars are designed not to let air or other materials in or out. Insulated means that heat cannot get in or out.

Evaporation indicates a phase change (vaporization) from liquid to gaseous phase; however, it does not imply that the matter has escaped its system.

A closed system can exchange energy (as heat or work) with its surroundings, but not matter.

An isolated system cannot exchange energy (as heat or work) or matter with the surroundings.

An open system can exchange energy and matter with the surroundings.

60. E is correct.

The *law of conservation of energy* states that the energy (e.g., potential or kinetic) of an isolated system remains constant and is conserved. Energy can be neither created nor destroyed but is transformed from one form to another. For instance, chemical energy can be converted to kinetic energy in the explosion of a firecracker.

The *law of conservation of mass* states that for any system closed to transfers of matter and energy, the mass of the system must remain constant over time.

Law of Definite Proportions (Proust's law) states that a chemical compound contains the same proportion of elements by mass. The law of definite proportions forms the basis of stoichiometry (i.e., from the known amounts of separate reactants, the amount of the product can be calculated).

Avogadro's law is an experimental gas law relating the volume of a gas to the amount of substance of gas present. It states that equal volumes of gases at the same temperature and pressure have the same number of molecules.

Boyle's law is an experimental gas law that describes how the pressure of a gas tends to increase as the volume of a gas decreases. It states that the absolute pressure exerted by a given mass of an ideal gas is inversely proportional to the volume it occupies if the temperature and amount of gas remain unchanged within a closed system.

Practice Set 4: Questions 61–80

---

61. C is correct.

ΔH refers to enthalpy (or heat).

Endothermic reactions have heat as a reactant.

Exothermic reactions have heat as a product.

Exothermic reactions release heat and cause the temperature of the immediate surroundings to rise (i.e., a net loss of energy).

Endothermic processes absorb heat and cool the surroundings (i.e., a net gain of energy).

Endothermic reactions absorb energy to break strong bonds to form a less stable state (i.e., positive enthalpy).

Exothermic reactions release energy when forming stronger bonds to produce a more stable state (i.e., negative enthalpy).

The reaction is nonspontaneous (i.e., endergonic) if the products are less stable than the reactants and ΔG is positive.

The reaction is spontaneous (i.e., exergonic) if the products are more stable than the reactants and ΔG is negative.

62. C is correct.

Temperature is a measure of the average kinetic energy of the molecules.

Kinetic energy is proportional to temperature.

In most thermodynamic equations, temperatures are expressed in Kelvin, so convert the Celsius temperatures to Kelvin:

$$25\ °C + 273.15 = 298.15\ K$$

$$50\ °C + 273.15 = 323.15\ K$$

Calculate the kinetic energy using simple proportions:

$$KE = (323.15\ K / 298.15\ K) \times 500\ J$$

$$KE = 540\ J$$

63. C is correct.

The equation relates to the change in entropy (ΔS) at different temperatures.

Entropy increases at higher temperatures (i.e., increased kinetic energy).

64. B is correct.

A change is exothermic when energy is released to the surroundings.

Energy loss occurs when a substance changes to a more rigid phase (e.g., gas → liquid, or liquid → solid).

65. A is correct.

Potential Energy **Stored energy and the energy of position (gravitational)**	**Kinetic Energy** **Energy of motion: motion of waves, electrons, atoms, molecules, and substances**
Chemical Energy Chemical energy is the energy stored in the bonds of atoms and molecules. Examples of stored chemical energy: biomass, petroleum, natural gas, propane, coal.	**Electrical Energy** Electrical energy is the movement of electrons. Lightning and electricity are examples of electrical energy.

66. B is correct.

At equilibrium, there is no potential, and therefore neither direction of the reaction is favored.

Reaction potential (E) indicates the tendency of a reaction to be spontaneous in either direction.

Because the solution is in equilibrium, no further reactions occur in either direction; $E = 0$.

67. D is correct.

The transformation from solid to gas is accompanied by increased entropy (i.e., disorder).

The molecules of gas have greater kinetic energy than the molecules of solids.

68. C is correct.

The reaction is nonspontaneous (i.e., endergonic) if the products are less stable than the reactants and ΔG is positive.

The reaction is spontaneous (i.e., exergonic) if the products are more stable than the reactants and ΔG is negative.

69. E is correct.

The spontaneity of a reaction is determined by evaluating the Gibbs free energy, or ΔG.

$$\Delta G = \Delta H - \mathrm{T}\Delta S$$

A reaction is spontaneous if $\Delta G < 0$ (or ΔG is negative).

For ΔG to be negative, ΔH has to be less than TΔS.

A reaction is nonspontaneous if $\Delta G > 0$ (or ΔG is positive and ΔH is greater than TΔS).

70. B is correct.

The reaction is nonspontaneous (i.e., endergonic) if the products are less stable than the reactants and ΔG is positive.

The reaction is spontaneous (i.e., exergonic) if the products are more stable than the reactants and ΔG is negative.

Entropy (ΔS) is determined by the relative number of molecules (compounds):

The reactants have one molecule (with a coefficient total of 2).

The products have two molecules (with a coefficient total of 7).

The number of molecules has increased from reactants to products. Therefore, the entropy (i.e., disorder) has increased.

71. D is correct.

ΔG indicates the available energy to do non P–V work.

This is a useful quantity for processes like batteries and living cells that do not have expanding gases.

72. C is correct.

The internal energy of a system is the energy contained within the system, excluding the kinetic energy of motion of the system as a whole and the potential energy of the system as a whole due to external force fields.

The internal energy of a system can be changed by transfers of matter and by work and heat transfer.

When impermeable container walls prevent matter transfer, the system is closed.

The First law of thermodynamics states that the increase in internal energy equals the heat added plus the work done on the system by its surroundings. If the container walls do not pass energy or matter, the system is isolated, and its internal energy remains constant.

Bond energy is internal potential energy (PE), while thermal energy is internal kinetic energy (KE).

73. D is correct.

Endothermic reactions absorb energy to break strong bonds to form a less stable state (i.e., positive enthalpy).

Exothermic reactions release energy when forming stronger bonds to produce a more stable state (i.e., negative enthalpy).

To predict spontaneity of reaction, use the Gibbs free energy equation:

$$\Delta G = \Delta H^\circ - \mathrm{T}\Delta S$$

The reaction is nonspontaneous (i.e., endergonic) if the products are less stable than the reactants and ΔG is positive.

The reaction is spontaneous (i.e., exergonic) if the products are more stable than the reactants and ΔG is negative.

It is the total amount of energy that is determinative.

Some bonds are stronger than others, so there is a net gain or net loss of energy when formed.

74. E is correct.

The starting material is ice at −25 °C, which means its temperature must be raised to 0 °C before it melts.

The first term in the heat calculation is:

specific heat of ice (A) × mass × change of temperature

75. C is correct.

The heat of formation of water vapor is similar to the highly exothermic process for the heat of hydrogen combustion.

Heat is required to convert H_2O (*l*) into vapor; therefore, the heat of formation of water vapor must be less exothermic than H_2O (*l*).

Solid → liquid → gas: endothermic because the reaction absorbs energy to break strong bonds to form a less stable state (i.e., positive enthalpy).

Gas → liquid → solid: exothermic because the reaction releases energy during the formation of stronger bonds to produce the more stable state (i.e., negative enthalpy).

76. E is correct.

77. B is correct.

Einstein established the equivalence of mass and energy, whereby the energy equivalent of mass is given by $E = mc^2$.

Energy can be converted to mass, and mass can be converted to energy as long as the total energy, including mass-energy, is conserved.

In nuclear fission, the mass of the parent nucleus is greater than the sum of the masses of the daughter nuclei. This deficit in mass is converted into energy.

78. D is correct.

A closed system can exchange energy (as heat or work) but not matter with its surroundings.

An isolated system cannot exchange energy (as heat or work) or matter with the surroundings.

An open system can exchange energy and matter with the surroundings.

79. A is correct.

Gases have the highest entropy (i.e., disorder or randomness).

Solutions have higher entropy than pure phases since the molecules are more scattered.

The liquid phase has less entropy than the aqueous phase.

80. A is correct.

In a chemical reaction, bonds within reactants are broken, and new bonds form.

Therefore, in bond dissociation problems:

$\Delta H_{reaction}$ = sum of bond energy in reactants – the sum of bond energy in products.

This is the opposite of ΔH_f problems, where:

$\Delta H_{reaction}$ = sum of ΔH_f product – sum of ΔH_f reactant

For $N_2 + 3\ H_2 \rightarrow 2\ NH_3$

$\Delta H_{reaction}$ = sum of bond energy in reactants – sum of bond energy in products

$\Delta H_{reaction} = [(N{\equiv}N) + 3(H–H)] – [2 \times 3(N–H)]$

$\Delta H_{reaction} = [(946\ kJ) + (3 \times 436\ kJ) – (2 \times (3 \times 389\ kJ)]$

$\Delta H_{reaction} = –80\ kJ$

Notes for active learning

Notes for active learning

APPENDIX

Periodic Table of the Elements

Key:

atomic number
Symbol
name
abridged standard atomic weight

1	2	3	4	5	6	7	8	9	10	11	12	13	14	15	16	17	18
1 **H** hydrogen 1.0080 ± 0.0002																	2 **He** helium 4.0026 ± 0.0001
3 **Li** lithium 6.94 ± 0.06	4 **Be** beryllium 9.0122 ± 0.0001											5 **B** boron 10.81 ± 0.02	6 **C** carbon 12.011 ± 0.002	7 **N** nitrogen 14.007 ± 0.001	8 **O** oxygen 15.999 ± 0.001	9 **F** fluorine 18.998 ± 0.001	10 **Ne** neon 20.180 ± 0.001
11 **Na** sodium 22.990 ± 0.001	12 **Mg** magnesium 24.305 ± 0.002											13 **Al** aluminium 26.982 ± 0.001	14 **Si** silicon 28.085 ± 0.001	15 **P** phosphorus 30.974 ± 0.001	16 **S** sulfur 32.06 ± 0.02	17 **Cl** chlorine 35.45 ± 0.01	18 **Ar** argon 39.95 ± 0.16
19 **K** potassium 39.098 ± 0.001	20 **Ca** calcium 40.078 ± 0.004	21 **Sc** scandium 44.956 ± 0.001	22 **Ti** titanium 47.867 ± 0.001	23 **V** vanadium 50.942 ± 0.001	24 **Cr** chromium 51.996 ± 0.001	25 **Mn** manganese 54.938 ± 0.001	26 **Fe** iron 55.845 ± 0.002	27 **Co** cobalt 58.933 ± 0.001	28 **Ni** nickel 58.693 ± 0.001	29 **Cu** copper 63.546 ± 0.003	30 **Zn** zinc 65.38 ± 0.02	31 **Ga** gallium 69.723 ± 0.001	32 **Ge** germanium 72.630 ± 0.008	33 **As** arsenic 74.922 ± 0.001	34 **Se** selenium 78.971 ± 0.008	35 **Br** bromine 79.904 ± 0.003	36 **Kr** krypton 83.798 ± 0.002
37 **Rb** rubidium 85.468 ± 0.001	38 **Sr** strontium 87.62 ± 0.01	39 **Y** yttrium 88.906 ± 0.001	40 **Zr** zirconium 91.224 ± 0.002	41 **Nb** niobium 92.906 ± 0.001	42 **Mo** molybdenum 95.95 ± 0.01	43 **Tc** technetium [97]	44 **Ru** ruthenium 101.07 ± 0.02	45 **Rh** rhodium 102.91 ± 0.01	46 **Pd** palladium 106.42 ± 0.01	47 **Ag** silver 107.87 ± 0.01	48 **Cd** cadmium 112.41 ± 0.01	49 **In** indium 114.82 ± 0.01	50 **Sn** tin 118.71 ± 0.01	51 **Sb** antimony 121.76 ± 0.01	52 **Te** tellurium 127.60 ± 0.03	53 **I** iodine 126.90 ± 0.01	54 **Xe** xenon 131.29 ± 0.01
55 **Cs** caesium 132.91 ± 0.01	56 **Ba** barium 137.33 ± 0.01	57-71 lanthanoids	72 **Hf** hafnium 178.49 ± 0.01	73 **Ta** tantalum 180.95 ± 0.01	74 **W** tungsten 183.84 ± 0.01	75 **Re** rhenium 186.21 ± 0.01	76 **Os** osmium 190.23 ± 0.03	77 **Ir** iridium 192.22 ± 0.01	78 **Pt** platinum 195.08 ± 0.02	79 **Au** gold 196.97 ± 0.01	80 **Hg** mercury 200.59 ± 0.01	81 **Tl** thallium 204.38 ± 0.01	82 **Pb** lead 207.2 ± 1.1	83 **Bi** bismuth 208.98 ± 0.01	84 **Po** polonium [209]	85 **At** astatine [210]	86 **Rn** radon [222]
87 **Fr** francium [223]	88 **Ra** radium [226]	89-103 actinoids	104 **Rf** rutherfordium [267]	105 **Db** dubnium [268]	106 **Sg** seaborgium [269]	107 **Bh** bohrium [270]	108 **Hs** hassium [269]	109 **Mt** meitnerium [277]	110 **Ds** darmstadtium [281]	111 **Rg** roentgenium [282]	112 **Cn** copernicium [285]	113 **Nh** nihonium [286]	114 **Fl** flerovium [290]	115 **Mc** moscovium [290]	116 **Lv** livermorium [293]	117 **Ts** tennessine [294]	118 **Og** oganesson [294]

57 **La** lanthanum 138.91 ± 0.01	58 **Ce** cerium 140.12 ± 0.01	59 **Pr** praseodymium 140.91 ± 0.01	60 **Nd** neodymium 144.24 ± 0.01	61 **Pm** promethium [145]	62 **Sm** samarium 150.36 ± 0.02	63 **Eu** europium 151.96 ± 0.01	64 **Gd** gadolinium 157.25 ± 0.03	65 **Tb** terbium 158.93 ± 0.01	66 **Dy** dysprosium 162.50 ± 0.01	67 **Ho** holmium 164.93 ± 0.01	68 **Er** erbium 167.26 ± 0.01	69 **Tm** thulium 168.93 ± 0.01	70 **Yb** ytterbium 173.05 ± 0.02	71 **Lu** lutetium 174.97 ± 0.01
89 **Ac** actinium [227]	90 **Th** thorium 232.04 ± 0.01	91 **Pa** protactinium 231.04 ± 0.01	92 **U** uranium 238.03 ± 0.01	93 **Np** neptunium [237]	94 **Pu** plutonium [244]	95 **Am** americium [243]	96 **Cm** curium [247]	97 **Bk** berkelium [247]	98 **Cf** californium [251]	99 **Es** einsteinium [252]	100 **Fm** fermium [257]	101 **Md** mendelevium [258]	102 **No** nobelium [259]	103 **Lr** lawrencium [262]

International Union of Pure and Applied Chemistry (IUPAC), 4 May 2022

Notes for active learning

Common Chemistry Equations

Throughout the test the following symbols have the definitions specified unless otherwise noted.

L, mL	= liter(s), milliliter(s)	mm Hg	= millimeters of mercury
g	= gram(s)	J, kJ	= joule(s), kilojoule(s)
nm	= nanometer(s)	V	= volt(s)
atm	= atmosphere(s)	mol	= mole(s)

ATOMIC STRUCTURE

$$E = h\nu$$
$$c = \lambda\nu$$

E = energy
ν = frequency
λ = wavelength

Planck's constant, $h = 6.626 \times 10^{-34}$ J s

Speed of light, $c = 2.998 \times 10^{8}$ m s^{-1}

Avogadro's number $= 6.022 \times 10^{23}$ mol^{-1}

Electron charge, $e = -1.602 \times 10^{-19}$ coulomb

EQUILIBRIUM

$$K_c = \frac{[C]^c[D]^d}{[A]^a[B]^b}, \text{ where } a\,A + b\,B \rightleftarrows c\,C + d\,D$$

$$K_p = \frac{(P_C)^c(P_D)^d}{(P_A)^a(P_B)^b}$$

$$K_a = \frac{[H^+][A^-]}{[HA]}$$

$$K_b = \frac{[OH^-][HB^+]}{[B]}$$

$$K_w = [H^+][OH^-] = 1.0 \times 10^{-14} \text{ at } 25°C$$
$$= K_a \times K_b$$

$$pH = -\log[H^+],\ pOH = -\log[OH^-]$$

$$14 = pH + pOH$$

$$pH = pK_a + \log\frac{[A^-]}{[HA]}$$

$$pK_a = -\log K_a,\ pK_b = -\log K_b$$

Equilibrium Constants

K_c (molar concentrations)
K_p (gas pressures)
K_a (weak acid)
K_b (weak base)
K_w (water)

KINETICS

$$\ln[A]_t - \ln[A]_0 = -kt$$

$$\frac{1}{[A]_t} - \frac{1}{[A]_0} = kt$$

$$t_{1/2} = \frac{0.693}{k}$$

k = rate constant
t = time
$t_{1/2}$ = half-life

GASES, LIQUIDS, AND SOLUTIONS

$$PV = nRT$$

$$P_A = P_{total} \times X_A, \text{ where } X_A = \frac{\text{moles A}}{\text{total moles}}$$

$$P_{total} = P_A + P_B + P_C + \ldots$$

$$n = \frac{m}{M}$$

$$K = {}^\circ C + 273$$

$$D = \frac{m}{V}$$

$$KE \text{ per molecule} = \frac{1}{2}mv^2$$

Molarity, M = moles of solute per liter of solution

$$A = abc$$

P = pressure
V = volume
T = temperature
n = number of moles
m = mass
M = molar mass
D = density
KE = kinetic energy
v = velocity
A = absorbance
a = molar absorptivity
b = path length
c = concentration

Gas constant, $R = 8.314 \text{ J mol}^{-1}\text{ K}^{-1}$
$= 0.08206 \text{ L atm mol}^{-1}\text{ K}^{-1}$
$= 62.36 \text{ L torr mol}^{-1}\text{ K}^{-1}$
1 atm = 760 mm Hg
= 760 torr
STP = 0.00°C and 10^5 Pa

THERMOCHEMISTRY/ ELECTROCHEMISTRY

$$q = mc\Delta T$$

$$\Delta S^\circ = \sum S^\circ \text{ products} - \sum S^\circ \text{ reactants}$$

$$\Delta H^\circ = \sum \Delta H_f^\circ \text{ products} - \sum \Delta H_f^\circ \text{ reactants}$$

$$\Delta G^\circ = \sum \Delta G_f^\circ \text{ products} - \sum \Delta G_f^\circ \text{ reactants}$$

$$\Delta G^\circ = \Delta H^\circ - T\Delta S^\circ$$
$$= -RT \ln K$$
$$= -nFE^\circ$$

$$I = \frac{q}{t}$$

q = heat
m = mass
c = specific heat capacity
T = temperature
S° = standard entropy
H° = standard enthalpy
G° = standard free energy
n = number of moles
E° = standard reduction potential
I = current (amperes)
q = charge (coulombs)
t = time (seconds)

Faraday's constant, F = 96,485 coulombs per mole of electrons

$$1 \text{ volt} = \frac{1 \text{ joule}}{1 \text{ coulomb}}$$

Annotated Glossary of Chemistry Terms

A

Absolute entropy (of a substance) – the increase in the entropy of a substance as it goes from a perfectly ordered crystalline form at 0 K (where its entropy is zero) to the temperature in question.

Absolute zero – the zero point on the absolute temperature scale; –273.15 °C or 0 K; theoretically, the temperature at which molecular motion ceases (i.e., the system does not emit or absorb energy, atoms at rest).

Absorption spectrum – spectrum associated with the absorption of electromagnetic radiation by atoms (or other species), resulting from transitions from lower to higher energy states.

Accuracy – the degree to which a value is close to the actual value; see *precision.*

Acid – a substance that produces H^+ (*aq*) ions in an aqueous solution and gives a pH of less than 7.0; strong acids ionize entirely or almost entirely in dilute aqueous solution; weak acids ionize only slightly. It turns litmus red.

Acid dissociation constant – an equilibrium constant for dissociating a weak acid.

Acid rain – rainwater with a pH of less than 5.7; caused by the gases NO_2 (vehicle exhaust fumes) and SO_2 (from burning fossil fuels) dissolving in the rain. It kills fish, wildlife, and trees and destroys buildings and lakes.

Acidic salt – contains an ionizable hydrogen atom; does not necessarily produce acidic solutions.

Actinides – the fifteen chemical elements that are between actinium (89) and lawrencium (103).

Activated complex – a structure formed by a collision between molecules in which new bonds form.

Activation energy – the amount of energy that reactants must absorb in their ground states to reach the transition state needed for a reaction to occur.

Active metal – a metal with low ionization energy that loses electrons readily to form cations.

Activity (of a component of ideal mixture) – a dimensionless quantity whose magnitude is equal to the molar concentration in an ideal solution; equal to partial pressure in an ideal gas mixture; 1 for pure solids or liquids.

Activity series – a listing of metals (and hydrogen) in order of decreasing activity.

Actual yield – the amount of a specified pure product obtained from a given reaction; see *theoretical yield.*

Addition reaction – a reaction in which two atoms or groups of atoms are added to a molecule, one on each side of a double or triple bond.

Adhesive forces – forces of attraction between a liquid and another surface.

Adsorption – the adhesion of a species onto the surfaces of particles.

Aeration – the mixing of air into a liquid or a solid.

Alcohol – a hydrocarbon derivative containing a ~OH group attached to a carbon atom, not in an aromatic ring.

Alkali metals – the elements of Group IA on the periodic table (e.g., Na, K, Rb).

Alkaline battery – a dry cell in which the electrolyte contains KOH.

Alkaline earth metals – group IIA metals on the periodic table; see *earth metals*.

Allomer – a substance that has a different composition from another but the same crystalline structure.

Allotropes – elements with different structures (therefore different forms), such as carbon (e.g., diamond, graphite, and fullerene).

Allotropic modifications (allotropes) – different forms of the same element in the same physical state.

Alloy – a mixture of metals. For example, bronze is an alloy formed from copper and tin.

Alloying – mixing metals with other substances (usually other metals) to modify their properties.

Alpha (α) particle – a helium nucleus; helium ion with 2+ charge; an assembly of two protons and two neutrons.

Amorphous solid – a non-crystalline solid with no well-defined ordered structure.

Ampere – unit of electrical current; one ampere equals one coulomb per second.

Amphiprotism – the ability of a substance to exhibit amphiprotic by accepting donated protons.

Amphoterism – the ability to react with both acids and bases; to act as either an acid or a base.

Amplitude – the maximum distance that medium particles carrying the wave move from their rest position.

Anion – a negative ion; an atom or group of atoms that has gained one or more electrons.

Anode – in a cathode ray tube, the positive electrode (electrode at which oxidation occurs); the positive side of a dry cell battery or a cell.

Antibonding orbital – a molecular orbital higher in energy than any of the atomic orbitals from which it is derived; lends instability to a molecule or ion when populated with electrons; denoted with star (*) superscript.

Artificial transmutation – an artificially induced nuclear reaction caused by the bombardment of a nucleus with subatomic particles or small nuclei.

Associated ions – short-lived species formed by the collision of dissolved ions of opposite charges.

Atmosphere – a unit of pressure; the pressure supports a column of mercury 760 mm high at 0 °C.

Atom – a chemical element in its smallest form; it comprises neutrons and protons within the nucleus and electrons circling the nucleus; it is the smallest part of an element that can exist.

Atomic mass unit (amu) – one-twelfth of the mass of an atom of the carbon-12 isotope; used for stating atomic and formula weights; known as a dalton.

Atomic number – represents an element corresponding with the number of protons within the nucleus; the number of protons in the nucleus of the atom.

Atomic orbital (*AO*) – a region or volume in space where the probability of finding electrons is highest.

Atomic radius – radius of an atom.

Atomic weight – weighted average of the masses of the constituent isotopes of an element; the relative masses of atoms of different elements.

Aufbau (or *building up*) principle – describes the order in which electrons fill orbitals in atoms.

Autoionization – an ionization reaction between identical molecules.

Avogadro's Law – equal volumes of gases contain the same number of molecules at the same temperature and pressure.

Avogadro's number (N_A) – the number (6.022×10^{23}) of atoms, molecules, or particles found in precisely 1 mole of a substance.

B

Background radiation – extraneous to an experiment; usually the low-level natural radiation from cosmic rays and trace radioactive substances present in the environment.

Band – a series of very closely spaced nearly continuous molecular orbitals that belong to the crystal as a whole.

Band of stability – band containing nonradioactive nuclides in a plot of neutrons *vs.* their atomic number.

Band theory of metals – the theory that accounts for the bonding and properties of metallic solids.

Barometer – a device used to measure the pressure in the atmosphere.

Base – a substance that produces ^{-}OH (*aq*) ions in an aqueous solution; accepts a proton and has a high pH; strongly soluble bases are soluble in water and are entirely dissociated; weak bases ionize only slightly; a typical example of a base is sodium hydroxide (NaOH). It turns litmus blue.

Basic anhydride – the oxide of a metal that reacts with water to form a base.

Basic salt – a salt containing an ionizable OH group.

Beta (β) particle – an electron emitted from the nucleus when a neutron decays to a proton and an electron.

Binary acid – a binary compound in which H is bonded to one or more electronegative nonmetals.

Binary compound – consists of two elements; it may be ionic or covalent.

Binding energy (nuclear binding energy) – the energy equivalent ($E = mc^2$) of the mass deficiency of an atom (where E is the energy in joules, m is the mass in kilograms, and c is the speed of light in m/s^2).

Boiling – the phase transition of a liquid vaporizing.

Boiling point – the temperature at which the vapor pressure of a liquid is equal to the applied pressure; the *condensation point*.

Boiling point elevation – the increase in the boiling point of a solvent caused by the dissolution of a nonvolatile solute.

Bomb calorimeter – a device used to measure the heat transfer between a system and its surroundings at constant volume.

Bond – the attraction and repulsion between atoms and molecules is a cornerstone of chemistry.

Bond energy – the amount of energy necessary to break one mole of bonds in a substance, dissociating the substance in its gaseous state into atoms of its elements in the gaseous state.

Bond order – half the number of electrons in bonding orbitals minus half the electrons in antibonding orbitals.

Bonding orbital – a molecular orbit lower in energy than any of the atomic orbitals from which it is derived; lends stability to a molecule or ion when populated with electrons.

Bonding pair – pair of electrons involved in a covalent bond.

Boron hydrides – binary compounds of boron and hydrogen.

Born-Haber cycle – a series of reactions (and the accompanying enthalpy changes) which, when summed, represent the hypothetical one-step reaction by which elements in their standard states are converted into crystals of ionic compounds (and the accompanying enthalpy changes).

Boyle's Law – at a constant temperature, the volume occupied by a definite mass of a gas is inversely proportional to the applied pressure.

Breeder reactor – a nuclear reactor that produces more fissionable nuclear fuel than it consumes.

Brønsted-Lowrey acid – a chemical species that donates a proton.

Brønsted-Lowrey base – a chemical species that accepts a proton.

Buffer solution – resists change in pH; contains either a weak acid and a soluble ionic salt of the acid or a weak base and a soluble ionic salt of the base.

Buret – a piece of volumetric glassware, usually graduated in 0.1 mL intervals, used to deliver solutions for titrations in a quantitative (drop-like) manner; also spelled *burette*.

C

Calorie – the amount of heat required to raise the temperature of one gram of water from 14.5 °C to 15.5 °C; 1 calorie = 4.184 joules.

Calorimeter – a device used to measure the heat transfer between a system and its surroundings.

Canal ray – a stream of positively charged particles (cations) that moves toward the negative electrode in cathode ray tubes; observed to pass through canals in the negative electrode.

Capillary – a tube having a very small inside diameter.

Capillary action – the drawing of a liquid up the inside of a small-bore tube when adhesive forces exceed cohesive forces; the depression of the surface of the liquid when cohesive forces exceed the adhesive forces.

Catalyst – a chemical compound used to change the rate (to speed or slow it) of a regenerated reaction (i.e., not consumed) at the end of the reaction.

Catenation – the bonding of atoms of the same element into chains or rings (i.e., the ability of an element to bond with itself).

Cathode – the electrode at which reduction occurs; in a cathode ray tube, the negative electrode.

Cathodic protection – protection of a metal (making a cathode) against corrosion by attaching it to a sacrificial anode of a more easily oxidized metal.

Cathode ray tube – a closed glass tube containing gas under low pressure, with electrodes near the ends and a luminescent screen near the positive electrode; produces cathode rays when a high voltage is applied.

Cation – a positive ion; an atom or group of atoms that has lost one or more electrons.

Cell potential – the potential difference, E_{cell}, between oxidation and reduction half-cells under nonstandard conditions; the force in a galvanic cell pulls electrons through a reducing agent to an oxidizing agent.

Central atom – an atom in a molecule or polyatomic ion bonded to more than one other atom.

Chain reaction – a reaction that, once initiated, sustains itself and expands; a reaction in which reactive species, such as radicals, are produced in more than one step, allowing these reactive species to propagate the chain reaction.

Charles' Law – at constant pressure, the volume occupied by a definite mass of gas is directly proportional to its absolute temperature.

Chemical bonds – the attractive forces holding atoms together in elements or compounds.

Chemical change – when one or more new substances are formed.

Chemical equation – description of a chemical reaction by placing the formulas of the reactants on the left of an arrow and the formulas of the products on the right.

Chemical equilibrium – a state of dynamic balance in which the rates of forward and reverse reactions are equal; there is no net change in concentrations of reactants or products while a system is at equilibrium.

Chemical kinetics – studies rates and mechanisms of chemical reactions and factors they depend on.

Chemical periodicity – the variations in properties of elements with their position in the periodic table.

Chemical reaction – the change of one or more substances into another or multiple substances.

Cloud chamber – a device for observing the paths of speeding particles as vapor molecules condense on them to form fog-like tracks.

Cobalt chloride paper – water test; water changes the color from blue to pink.

Coefficient of expansion – the ratio of the change in the length or the volume of a body to the original length or volume for a unit change in temperature.

Cohesive forces – the forces of attraction among particles of a liquid.

Colligative properties – physical properties of solutions that depend upon the number but not the kind of solute particles present.

Collision theory – the theory of reaction rates that states that effective collisions between reactant molecules must occur for the reaction to occur.

Colloid – a heterogeneous mixture in which solute-like particles do not settle (e.g., many kinds of milk).

Combination reaction – two substances (elements or compounds) combine to form one compound.

Combustible – classification of liquid substances that burn based on flashpoints; any liquid having a flashpoint at or above 37.8 °C (100 °F) but below 93.3 °C (200 °F), except any mixture having components with flashpoints of 93.3 °C (200 °F) or higher, the total of which makes up 99% or more of the volume of the mixture.

Combustion (or *burning*) – an exothermic reaction between an oxidant and fuel with heat and often light.

Common ion effect – suppression of ionization of a weak electrolyte by the presence in the same solution of a strong electrolyte containing one of the same ions as the weak electrolyte.

Complex ions – ions resulting from coordinating covalent bonds between simple ions and other ions or molecules.

Composition stoichiometry – describes the quantitative (mass) relationships among elements in compounds.

Compound – a substance of two or more chemically bonded elements in fixed proportions; can be decomposed into constituent elements.

Compressed gas – a single or mixture of gases having (in a container) an absolute pressure exceeding 40 psi at 21.1 °C (70 °F).

Compression – an area in a longitudinal wave where the particles are closer and pushed in.

Concentration – the amount of solute per unit volume, the mass of solvent or solution.

Condensation – the phase change from gas to liquid.

Condensed phases – the liquid and solid phases; phases in which particles interact strongly.

Condensed states – the solid and liquid states.

Conduction – heat transfer between substances in direct contact with each other (i.e., must be touching); when particles of a hotter substance vibrate, these molecules bump into nearby particles and transfer some energy.

Conduction band – a vacant or partially filled band of energy levels just higher in energy than a filled band; a band within which, or into which, electrons must be promoted to allow electrical conduction to occur in a solid.

Conductor – a material that allows electric flow more freely.

Conjugate acid-base pair – in Brønsted-Lowry terms, reactant and product that differ by a proton, H^+.

Conformations – structures of a compound that differ by the extent of rotation about a single bond.

Continuous spectrum – contains wavelengths in a specified region of the electromagnetic spectrum.

Control rods – rods of materials such as cadmium or boron steel that act as neutron absorbers (not merely moderators), used in nuclear reactors to control neutron fluxes and therefore fission rates.

Conjugated double bonds – double bonds separated from each other by one single bond –C=C–C=C–

Contact process – the industrial process for sulfur trioxide (SO_3) and sulfuric acid (H_2SO_4) production from sulfur dioxide (SO_2).

Convection – the physical flow of matter when heat flows by energized molecules from one place to another through the movement of fluids. Transfer of heat through a liquid or a gas occurs when molecules of the liquid or gas move and carry the heat.

Coordinate covalent bond – a covalent bond with shared electrons furnished by the same species; a bond between a Lewis acid and a Lewis base.

Coordination compound or complex – a compound containing coordinate covalent bonds.

Coordination number – the number of donor atoms coordinated to a metal; the number of nearest neighbors of an atom or ion in describing crystals.

Coordination sphere – the metal ion and its coordinating ligands, but no uncoordinated counter-ions.

Corrosion – oxidation of metals (e.g., rusting) in the presence of air and moisture.

Coulomb – the SI unit of electrical charge; unit symbol – C.

Covalent bond – a force of attraction (chemical bond) formed by the sharing of electron pairs between two atoms.

Covalent compounds – compounds made of two or more nonmetal atoms bonded by sharing valence electrons.

Critical mass – the minimum mass of a particular fissionable nuclide in a given volume required to sustain a nuclear chain reaction.

Critical point – the combination of critical temperature and critical pressure of a substance.

Critical pressure – the pressure required to liquefy a gas (vapor) at its *Critical temperature*.

Critical temperature – the temperature above which a gas cannot be liquefied; the temperature above which a substance cannot exhibit distinct gas and liquid phases.

Crystal – a solid packed with ions, molecules, or atoms in an orderly fashion.

Crystal field stabilization energy – a measure of the net energy of stabilization gained by a metal ion's nonbonding d electrons due to complex formation.

Crystal field theory – bonding in transition metal complexes in which ligands and metal ions are treated as point charges; a purely ionic model; ligand point charges represent the crystal (electrical) field perturbing the metal's d orbitals containing nonbonding electrons.

Crystal lattice – a pattern of arrangement of particles in a crystal.

Crystal lattice energy – the amount of energy that holds a crystal together; the energy change when a mole of solid forms from its constituent molecules or ions (for ionic compounds) in their gaseous state (consistently negative).

Crystalline solid – a solid characterized by a regular, ordered arrangement of particles.

Curie (Ci) – the basic unit to describe the intensity of radioactivity in a sample of material; one curie equals 37 billion disintegrations per second or approximately the amount of radioactivity given off by 1 gram of radium.

Current – a flow of charged particles, such as electrons or ions, moving through an electrical space or conductor. It is measured as the net rate of flow of electric charge; the unit is an Ampere (A).

Cuvette – glassware used in spectroscopic experiments; usually made of plastic, glass, or quartz, and should be as clean and transparent as possible.

Cyclotron – a device for accelerating charged particles along a spiral path.

D

Dalton's Law (or the *law of partial pressures*) – the pressure exerted by a mixture of gases is the sum of the partial pressures of the individual gases.

Daughter nuclide – nuclide produced in nuclear decay.

Debye – the unit used to express dipole moments.

Degenerate – in orbitals, describes orbitals of the same energy.

Deionization – the removal of ions; in the case of water, mineral ions such as sodium, iron, and calcium.

Deliquescence – substances that absorb water from the atmosphere to form liquid solutions.

Delocalization – in reference to electrons, bonding electrons distributed among more than two atoms bonded; occurs in species that exhibit resonance.

Density – mass per unit volume; $D = M \times V$.

Deposition – settling particles within a solution; the direct solidification of vapor by cooling; see *sublimation*.

Derivative – a compound that can be imagined arising from a parent compound by replacing one atom with another atom or group of atoms; used extensively in organic chemistry to identify compounds.

Detergent – a soap-like emulsifier with sulfate, SO_3, or a phosphate group instead of carboxylate group.

Deuterium – an isotope of hydrogen whose atoms are twice as massive as ordinary hydrogen; deuterium atoms contain a proton and a neutron in the nucleus.

Dextrorotatory – refers to an optically active substance that rotates plane-polarized light clockwise, also known as *dextro*.

Diagonal similarities – chemical similarities in the Periodic Table of Elements of Period 2 to elements of Period 3 one group to the right, especially evident toward the left of the periodic table.

Diamagnetism – weak repulsion by a magnetic field.

Differential Scanning Calorimetry (DSC) – a technique for measuring temperature, direction, and magnitude of thermal transitions in a sample material by heating/cooling and comparing the amount of energy required to maintain its rate of temperature increase or decrease with an inert reference material under similar conditions.

Differential Thermal Analysis (DTA) – a technique for observing the temperature, direction, and magnitude of thermally induced transitions in a material by heating/cooling a sample and comparing its temperature with an inert reference material under similar conditions.

Differential thermometer – a thermometer used to measure minimal temperature changes accurately.

Dilution – the process of reducing the concentration of a solute in a solution, usually by mixing with more solvent.

Dimer – a molecule formed by combining two smaller (identical) molecules.

Dipole – electric or magnetic separation of charge; charge separation between two covalently bonded atoms.

Dipole-dipole interactions – attractive electrostatic forces between polar molecules (i.e., between molecules with permanent dipoles).

Dipole moment – the product of the distance separating opposite charges of equal magnitude; a measure of the polarity of a bond or molecule; a measured dipole refers to the dipole moment of an entire molecule.

Dispersing medium – the solvent-like phase in a colloid.

Dispersed phase – the solute-like species in a colloid.

Displacement reactions – reactions in which one element displaces another from a compound.

Disproportionation reactions – redox reactions in which the oxidizing agent and the reducing agent are the same species.

Dissociation – in an aqueous solution, the process by which a solid ionic compound separates into its ions.

Dissociation constant – equilibrium constant for dissociating a complex ion into a simple ion and coordinating species (ligands).

Dissolution or solvation – the spread of ions in a monosaccharide.

Distilland – the material in a distillation apparatus that is to be distilled.

Distillate – the material in a distillation apparatus collected in the receiver.

Distillation – separating a liquid mixture into its components based on differences in boiling points; the process in which components of a mixture are separated by boiling away the more volatile liquid; the vaporization of a liquid by heating and then the condensation of the vapor by cooling.

Domain – a cluster of atoms in a ferromagnetic substance, which align in the same direction in the presence of an external magnetic field.

Donor atom – a ligand atom whose electrons are shared with a Lewis acid.

***d*-orbitals** – beginning in the third energy level, a set of five degenerate orbitals per energy level, higher in energy than *s* and *p* orbitals of the same energy level.

Dosimeter – a small, calibrated electroscope worn by laboratory personnel to measure incident ionizing radiation or chemical exposure.

Double bond – covalent bond resulting from the sharing of four electrons (two pairs) between two atoms.

Double salt – a solid consisting of two co-crystallized salts.

Doublet – two peaks or bands of about equal intensity appearing close on a spectrogram.

Downs cell – an electrolytic cell for the commercial electrolysis of molten sodium chloride.

DP number – the degree of polymerization; the average number of monomer units per polymer unit.

Dry cells (voltaic cells) – ordinary batteries for appliances (e.g., flashlights, radios).

Dumas method – a method used to determine the molecular weights of volatile liquids.

Dynamic equilibrium – an equilibrium in which the processes occur continuously with no net change.

E

Earth metal – highly reactive elements in group IIA of the periodic table (includes beryllium, magnesium, calcium, strontium, barium, and radium); see *alkaline earth metal.*

Effective collisions – a collision between molecules resulting in a reaction, one in which the molecules collide with proper relative orientations and sufficient energy to react.

Effective molality – the sum of the molalities of solute particles in a solution.

Effective nuclear charge – the nuclear charge experienced by the outermost electrons of an atom; the actual nuclear charge minus the effects of shielding due to inner-shell electrons (e.g., a set of $d_{x2\text{-}y2}$ and d_{z2} orbitals); those d orbitals within a set with lobes directed along the *x*, *y*, and *z* axes.

Electrical conductivity – the measure of how easily an electric current can flow through a substance.

Electric charge – a measured property (coulombs) that determines electromagnetic interaction.

Electrochemical cell – using a chemical reaction, an electromotive force is generated.

Electrochemistry – the study of chemical changes produced by electrical current and electricity production by chemical reactions.

Electrodes – surfaces upon which oxidation and reduction half-reactions occur in electrochemical cells; a conductor dips into an electrolyte and allows the electrons to flow to and from the electrolyte.

Electrode potentials – potentials, E, of half-reactions as reductions vs. standard hydrogen electrode.

Electrolysis – 1) occurs in electrolytic cells; chemical decomposition occurs by passing an electric current through a solution containing ions. 2) producing a chemical change using electricity; used to split up water into H and O_2.

Electrolyte – a substance (i.e., anions, cations) which, when dissolved in water, conducts electricity. An ionic solution that conducts a certain amount of current is categorized into two types: weak and strong.

Electrolytic cells – electrochemical cells in which electrical energy causes nonspontaneous redox reactions to occur (i.e., forced to occur by applying an outside source of electrical energy).

Electrolytic conduction – electrical current passes through ions in a solution or pure liquid.

Electromagnetic radiation – energy propagated using electric and magnetic fields that oscillate in directions perpendicular to the direction of travel of the energy; a type of wave that can go through vacuums as well as material; classified as a "self-propagating wave."

Electromagnetism – fields of an electric charge and electric properties that change how particles move and interact.

Electromotive force – a device that gains energy as electric charges pass through it.

Electromotive series – the relative order of tendencies for elements and their simple ions to act as oxidizing or reducing agents; also known as the "activity series."

Electron – a subatomic particle having a mass of 0.00054858 amu and a charge of –1.

Electron affinity – the energy absorbed in the process in which an electron is added to a neutral isolated gaseous atom to form a gaseous ion with a 1– charge; it has a negative value if energy is released.

Electron configuration – the specific distribution of electrons in atomic orbitals of atoms or ions.

Electron-deficient compounds – at least one atom (other than H) shares fewer than eight electrons.

Electron shells – an orbital around an atom's nucleus with a fixed number of electrons (usually 2 or 8).

Electronic transition – the transfer of an electron from one energy level to another.

Electronegativity – a measure of the relative tendency of an atom to attract electrons to itself when chemically combined with another atom.

Electronic geometry – the geometric arrangement of orbitals containing the shared and unshared electron pairs surrounding the central atom or polyatomic ion.

Electrophile – positively charged or electron-deficient.

Electrophoresis – separating ions by migration rate and direction of migration in an electric field.

Electroplating – a metal is covered with another metal layer using electricity; plating a metal onto a (cathodic) surface by electrolysis.

Element – a substance that cannot be decomposed into simpler substances by chemical means; defined by its *atomic number*. A substance that cannot be split into simpler substances by chemical means.

Eluant (or eluent) – the solvent used in the process of elution, as in liquid chromatography.

Eluate – a solvent (or mobile phase) which passes through a chromatographic column and removes the sample components from the stationary phase.

Emission spectrum – the emission of electromagnetic radiation by atoms (or other species) resulting from electronic transitions from higher to lower energy states.

Empirical formula – the simplest whole-number ratio of atoms of each element present in a compound; also known as the *simplest formula*.

Emulsifying agent – a substance that coats the particles of the dispersed phase and prevents coagulation of colloidal particles; an emulsifier.

Emulsion – colloidal suspension of a liquid in a liquid.

Endothermic – describes processes that absorb heat energy (H).

Endothermicity – the absorption of heat by a system as the process occurs.

Endpoint – the point at which an indicator changes color and titration stops.

Energy – a system's ability to do work.

Enthalpy (*H*) – the heat content of a specific amount of substance; *E*= PV.

Entropy (*S*) – a thermodynamic state or property that measures the degree of disorder (i.e., randomness) of a system; the amount of energy not available for work in a closed thermodynamic system (usually denoted by S).

Enzyme – a protein that acts as a catalyst in biological systems.

Equation of state – an expression that describes the behavior of matter in a given state; the van der Waals equation describes the behavior of the gaseous state.

Equilibrium or chemical equilibrium – a state of dynamic balance with the rates of forward and reverse reactions equal; the state of a system where neither forward nor reverse reaction is thermodynamically favored.

Equilibrium constant – a quantity characterizing equilibrium position for a reversible reaction; its magnitude is equal to mass action expression at equilibrium; equilibrium, "K," varies with temperature.

Equivalence point – the point when chemically equivalent amounts of reactants have reacted.

Equivalent weight – an oxidizing or reducing agent whose mass gains (oxidizing agents) or loses (reducing agents) 6.022×10^{23} electrons in a redox reaction.

Evaporation – vaporization of a liquid below its boiling point.

Evaporation rate – the rate at which a particular substance vaporizes (evaporate) compared to the rate of a known substance, such as ethyl ether, especially useful for health and fire-hazard considerations.

Excited state – any state other than the ground state of an atom or molecule; see *ground state*.

Exothermic – describes processes that release heat energy (*H*).

Exothermicity – the release of heat by a system as a process occurs.

Explosive – a chemical or compound that causes a sudden, almost instantaneous release of pressure, gas, heat, and light when subjected to sudden shock, pressure, high temperature, or applied potential.

Explosive limits – the range of concentrations over which a flammable vapor mixed with the proper ratios of air will ignite or explode if a source of ignition is provided.

Extensive property – a property that depends upon the amount of material in a sample.

Extrapolate – to estimate the value of a result outside the range of a series of known values; a technique used in the standard additions calibration procedure.

F

Faraday constant – a unit of electrical charge widely used in electrochemistry and equal to ~ 96,500 coulombs; represents 1 mole of electrons, or the Avogadro number of electrons: 6.022×10^{23} electrons.

Faraday's law of electrolysis – a two-part law that Michael Faraday published about electrolysis. 1. the mass of a substance altered at an electrode during electrolysis is directly proportional to the quantity of electricity transferred at that electrode. 2. the mass of an elemental material altered at an electrode is directly proportional to the element's equivalent weight; one equivalent weight of a substance is produced at each electrode during the passage of 96,487 coulombs of charge through an electrolytic cell.

Fast neutron – a neutron ejected at high kinetic energy in a nuclear reaction.

Ferromagnetism – the ability of a substance to become permanently magnetized by exposure to an external magnetic field.

Flashpoint – the temperature at which a liquid yields enough flammable vapor to ignite; various recognized industrial testing methods exist; therefore, the method used must be specified.

Fluorescence – absorption of high-energy radiation and subsequent emission of visible light.

First Law of Thermodynamics – the amount of energy in the universe is constant (i.e., energy is neither created nor destroyed in ordinary chemical reactions and physical changes); known as the Law of Conservation of Energy.

Fluids – substances that flow freely; gases and liquids.

Flux – a substance added to react with the charge or a product of its reduction; in metallurgy, it is usually added to lower the melting point.

Foam – colloidal suspension of a gas in a liquid.

Formal charge – a method of counting electrons in a covalently bonded molecule or ion; it counts bonding electrons as though they were equally shared between the two atoms.

Formula – a combination of symbols that indicates the chemical composition of a substance.

Formula unit – the smallest repeating unit of a substance; the molecule for nonionic substances.

Formula weight – the mass of one formula unit of a substance in atomic mass units.

Fossil fuels – formed from the remains of plants and animals that lived millions of years ago.

Fractional distillation – when a fractioning column is used in a distillation apparatus to separate the components of a liquid mixture with different boiling points.

Fractional precipitation – removal of some ions from a solution by precipitation while leaving other ions with similar properties in the solution.

Free energy change – the indicator of the spontaneity of a process at constant T and P (e.g., if ΔG is negative, the process is spontaneous).

Free radical – a highly reactive chemical species carrying no charge and having a single unpaired electron in an orbital.

Freezing – phase transition from liquid to solid.

Freezing point depression – the decrease in the freezing point of a solvent caused by the presence of a solute.

Frequency – the number of repeating points on a wave that passes a given observation point per unit time; the unit is 1 hertz = 1 cycle per 1 second.

Fuel – any substance that burns in oxygen to produce heat.

Fuel cells – a voltaic cell that converts the chemical energy of a fuel and an oxidizing agent directly into electrical energy continuously.

G

Gamma (γ) ray – a highly penetrating type of nuclear radiation similar to X-ray radiation, except that it comes from within the nucleus of an atom and has higher energy; energy-wise, very similar to cosmic rays, except that cosmic rays originate from outer space.

Galvanic cell – a battery made from an electrochemical cell with two metals connected by a salt bridge.

Galvanizing – placing a thin layer of zinc on a ferrous material to protect the underlying surface from corrosion.

Gangue – sand, rock, and other impurities surrounding the mineral of interest in an ore.

Gas – a state of matter in which the particles have no definite shape or volume, though they fill their container.

Gay-Lussac's Law – the expression used for each of the two relationships named after the French chemist Joseph Louis Gay-Lussac concerning the properties of gases; more usually applied to his law of combining volumes.

Geiger counter – a gas-filled tube that discharges electrically when ionizing radiation passes through it.

Gel – colloidal suspension of a solid dispersed in a liquid; a semi-rigid solid.

Gibbs (free) energy – the thermodynamic state function of a system that indicates the amount of energy available for the system to do useful work at constant T and P; a value that indicates the spontaneity of a reaction (usually denoted by *G*).

Graham's Law – the rates of effusion of gases are inversely proportional to the square roots of their molecular weights or densities.

Ground state – the lowest energy state or most stable state of an atom, molecule, or ion; see *excited state*.

Group – a vertical column in the periodic table; known as a family.

H

Haber process – a process for the catalyzed industrial production of ammonia from N_2 and H_2 at high temperature and pressure.

Half-cell – the compartment in which the oxidation or reduction half-reaction occurs in a voltaic cell.

Half-life – the time required for half of a reactant to be converted into product(s); the time required for half of a given sample to undergo radioactive decay.

Half-reaction – the oxidation or the reduction part of a redox reaction.

Halogens – group VIIA elements: F, Cl, Br, I; halogens are nonmetals.

Hard water – water high in dissolved minerals that makes it difficult to form lather with soap.

Heat – a form of energy that flows between two samples of matter because of temperature differences.

Heat capacity – the amount of heat required to raise the temperature of a mass one degree Celsius.

Heat of condensation – the amount of heat that must be removed from one gram of vapor at its condensation point to condense the vapor with no change in temperature.

Heat of crystallization – the amount of heat that must be removed from one gram of a liquid at its freezing point to freeze it with no change in temperature.

Heat of fusion – the amount of heat required to melt one gram of a solid at its melting point with no change in temperature; usually expressed in J/g; the molar heat of fusion is the amount of heat required to melt one mole of a solid at its melting point with no change in temperature and is usually expressed in kJ/mol.

Heat of solution – the amount of heat absorbed in forming a solution that contains one mole of the solute; the value is positive if heat is absorbed (endothermic) and negative if heat is released (exothermic).

Heat of vaporization – the amount of heat required to vaporize one gram of a liquid at its boiling point with no change in temperature; usually expressed in J/g; the molar heat of vaporization is the amount of heat required to vaporize one mole of liquid at its boiling point with no change in temperature and is usually expressed as ion kJ/mol.

Heisenberg uncertainty principle – states that it is impossible to determine the momentum and position of an electron simultaneously with absolute accuracy.

Henry's Law – the gas pressure above a solution is proportional to concentration of the gas in solution.

Hess' Law of heat summation – the enthalpy change for a reaction is the same whether it occurs in one step or a series of steps.

Heterogeneous catalyst – exists in a different phase (solid, liquid, or gas) from the reactants; a contact catalyst.

Heterogeneous equilibria – equilibria involving species in more than one phase.

Heterogeneous mixture – a mixture that does not have uniform composition and properties throughout.

Heteronuclear – consisting of different elements.

High spin complex – crystal field designation for an outer orbital complex; t_{2g} and e_g orbitals are singly occupied before pairing occurs.

Homogeneous catalyst – in the same phase (solid, liquid, or gas) as the reactants.

Homogeneous equilibria – when *reagents* and *products* are in the same phase (gases, liquids, or solids).

Homogeneous mixture – a mixture which has uniform composition and properties throughout.

Homologous series – compounds with each member differing from the next by a specific number and kind of atoms.

Homonuclear – consisting of only one element.

Hund's rule – single electrons must occupy orbitals of a given sublevel before pairing begins; see *Aufbau* (or *building up*) *principle*.

Hybridization – mixing atomic orbitals to form a new set of atomic orbitals with the same electron capacity and properties and energies intermediate between the original unhybridized orbitals.

Hydrate – a solid compound that contains a definite percentage of bound water.

Hydrate isomers – crystalline complexes that differ in whether water exists inside or outside the coordination sphere.

Hydration – the reaction of a substance with water.

Hydration energy – the energy change accompanying the hydration of a mole of gas and ions.

Hydride – a binary compound of hydrogen.

Hydrocarbons – compounds that contain only carbon and hydrogen.

Hydrogen bond – a relatively strong dipole-dipole interaction (but still considerably weaker than the covalent or ionic bonds) between molecules containing hydrogen directly bonded to a small, highly electronegative atom, such as N, O or F.

Hydrogenation – the reaction in which hydrogen adds across a double or triple bond.

Hydrogen-oxygen fuel cell – hydrogen is the fuel (reducing agent), and oxygen is the oxidizing agent.

Hydrolysis – the reaction of a substance with water or its ions.

Hydrolysis constant – an equilibrium constant for a hydrolysis reaction.

Hydrometer – a device used to measure the densities of liquids and solutions.

Hydrophilic colloids – colloidal particles that repel water molecules.

I

Ideal gas – a hypothetical gas that obeys the postulates of the Kinetic Molecular Theory.

Ideal gas law – the product of pressure and the volume of an ideal gas is directly proportional to the number of moles of the gas and the absolute temperature. $PV = nRT$

Ideal solution – obeys Raoult's Law strictly.

Immiscible liquids – do not mix to form a solution (e.g., oil and water).

Indicators – for acid-base titrations, organic compounds that exhibit different colors in solutions of different acidities, determine the point at which the reaction between two solutes is complete.

Inert pair effect – characteristic of the post-transition minerals; the tendency of the electrons in the outermost atomic *s* orbital to remain un-ionized or unshared in compounds of post-transition metals.

Inhibitory catalyst – an inhibitor; a catalyst that decreases the rate of reaction.

Inner orbital complex – valence bond designation for a complex in which the metal ion utilizes d orbitals for one shell inside the outermost occupied shell in its hybridization.

Inorganic chemistry – a part of chemistry concerned with inorganic (non-carbon-based) compounds.

Insulator – a material that resists the flow of electric current or heat transfer; it does not allow heat to flow easily.

Insoluble compound – a substance that will not dissolve in a solvent, even after mixing.

Integrated rate equation – an expression giving the concentration of a reactant remaining after a specified time; it has a different mathematical form for different orders of reactants.

Intermolecular forces – forces between individual particles (atoms, molecules, ions) of a substance.

Ion – a molecule that has gained or lost electrons; an atom or a group of atoms carries an electric charge (Na^+).

Ion product for water – equilibrium constant for water ionization; $Kw = [H_3O^+]\cdot[^-OH] = 1.00 \times 10^{-14}$ at 25 °C.

Ionic bond – the electrostatic attraction between oppositely charged ions, resulting from a transfer of electrons.

Ionic bonding – chemical bonding resulting from transferring electrons from one atom or group.

Ionic compounds – compounds containing predominantly ionic bonding.

Ionic geometry – arrangement of atoms (not lone pairs of electrons) about the central atom of a polyatomic ion.

Ionization – the breaking up of a compound into separate ions; in an aqueous solution, the process by which a molecular compound reacts with water and forms ions.

Ionization constant – equilibrium constant for the ionization of a weak electrolyte.

Ionization energy – the minimum amount of energy required to remove the most loosely held electron of an isolated gaseous atom or ion.

Ion exchange – a method of removing hardness from water, it replaces the positive ions that cause the hardness with H^+ ions.

Ionization isomers – result from the interchange of ions inside and outside the coordination sphere.

Isoelectric – having the same electronic configurations.

Isomers – different substances with the same formula.

Isomorphous – refers to crystals having the same atomic arrangement.

Isotopes – two or more forms of atoms of the same element with different masses; atoms containing the same number of protons but different numbers of neutrons.

IUPAC – acronym for "International Union of Pure and Applied Chemistry."

J

Joule (J) – a unit of energy in the SI system; one joule is 1 $kg\cdot m^2/s^2$, which is 0.2390 calories.

K

K capture – absorption of a *K* shell (n = 1) electron by a proton as it is converted to a neutron.

Kelvin – a unit of measure for temperature based upon an absolute scale.

Kinetics – a subfield of chemistry specializing in reaction rates.

Kinetic energy (*KE*) – energy that matter processes through its motion.

Kinetic Molecular Theory – a theory that attempts to explain macroscopic observations on gases in microscopic or molecular terms.

L

Lanthanides – elements 57 (lanthanum) through 71 (lutetium); grouped because of their similar behavior in chemical reactions.

Lanthanide contraction – a decrease in the radii of the elements following the lanthanides compared to what would be expected if there were no *f*-transition metals.

Latent heat – the energy absorbed or released when a substance changes state without changing temperature.

Lattice – a unique arrangement of atoms or molecules in a crystalline liquid or solid.

Law of combining volumes (Gay-Lussac's Law) – at constant temperature and pressure, the volumes of reacting gases (and any gaseous products) can be expressed as ratios of small whole numbers.

Law of conservation of energy – energy cannot be created or destroyed; it can only change form.

Law of conservation of matter – there is no detectable change in the quantity of matter during an ordinary chemical reaction.

Law of conservation of matter and energy – the amount of matter and energy in the universe is fixed.

Law of definite proportions (law of constant composition) – different samples of a pure compound contain the same elements in the same proportions by mass.

Law of partial pressures (or *Dalton's Law*) – the pressure exerted by a mixture of gases is the sum of the partial pressures of the individual gases.

Laws of thermodynamics – physical laws which define quantities of thermodynamic systems describe how they behave and (by extension) set certain limitations, such as perpetual motion.

Lead storage battery – a secondary voltaic cell used in most automobiles.

Leclanche cell – a common type of *dry cell*.

Le Chatelier's principle – states that a system at equilibrium, or striving to attain equilibrium, responds in such a way as to counteract any stress placed upon it; if stress (change of conditions) is applied to a system at equilibrium, the system will shift in the direction that reduces stress.

Leveling effect – acids stronger than the acid characteristic of the solvent react with the solvent to produce that acid; a similar statement applies to bases. The strongest acid (base) that can exist in a given solvent is the acid (base) characteristic of the solvent.

Levorotatory (or *levo*) – an optically active substance rotates plane-polarized light counterclockwise.

Lewis acid – any species that can accept a share in an electron pair.

Lewis base – any species that can make available a pair of electrons.

Lewis dot formula (electron dot formula) – representation of a molecule, ion, or formula unit by showing atomic symbols and only outer shell electrons.

Ligand – a Lewis base in a coordination compound.

Light – that portion of the electromagnetic spectrum visible to the naked eye; known as "visible light."

Limiting reactant – a substance that stoichiometrically limits the number of products formed.

Linear accelerator – a device used for accelerating charged particles along a straight line path.

Line spectrum – an atomic emission or absorption spectrum.

Linkage isomers – a particular ligand bonds to a metal ion through different donor atoms.

Liquid – a state of matter which takes the shape of its container.

Liquid aerosol – colloidal suspension of a liquid in a gas.

London dispersion forces – very weak and very short-range attractive forces between short-lived temporary (induced) dipoles; known as "dispersion forces."

Lone pair – a pair of electrons residing on one atom and not shared by other atoms; an unshared pair.

Low spin complex – crystal field designation for an inner orbital complex; contains electrons paired t_{2g} orbitals before e_g orbitals are occupied in octahedral complexes.

Lubricant – a substance capable of reducing friction (i.e., a force that opposes the direction of motion).

M

Magnetic field – a space around a magnet where magnetism can be detected.

Magnetic quantum number – quantum mechanical solution to a wave equation designating the orbital within a given set (*s*, *p*, *d*, *f*) in which an electron resides.

Manometer – a two-armed barometer.

Mass (*m*) – a measure of the amount of matter in an object; mass is usually measured in grams or kilograms.

Mass action expression – for a reversible reaction, aA + bB cC + dD; the product of the concentrations of the products (species on the right), each raised to the power corresponding to its coefficient in the balanced chemical equation, divided by the product of the concentrations of reactants (species on the left), each raised to the power corresponding to its coefficient in the balanced equation; at equilibrium the mass action expression is equal to *K*.

Mass deficiency – the amount of matter converted into energy when an atom forms from constituent particles.

Mass number – the sum of the numbers of protons and neutrons in an atom; an integer.

Mass spectrometer – an instrument that measures the charge-to-mass ratio of charged particles.

Matter – anything that has mass and occupies space.

Mechanism – the sequence of steps by which reactants are converted into products.

Melting point – the temperature at which liquid and solid coexist in equilibrium.

Meniscus – the shape assumed by the surface of a liquid in a cylindrical container.

Melting – the phase change from a solid to a liquid.

Metal – a chemical element that is a good conductor of electricity and heat and forms cations and ionic bonds with nonmetals; elements below and to the left of the stepwise division (metalloids) in the upper right corner of the periodic table; about 80% of known elements are metals.

Metallic bonding – bonding within metals due to the electrical attraction of positively charged metal ions for mobile electrons that belong to the crystal.

Metallic conduction – conduction of electrical current through a metal or along a metallic surface.

Metalloid – a substance with the properties of metals and nonmetals (B, Al, Si, Ge, As, Sb, Te, Po and At).

Metathesis reactions – reactions in which two compounds react to form two new compounds, with no changes in oxidation number; reactions in which the ions of two compounds exchange partners.

Method of initial rates – method of determining the rate-law expression by carrying out a reaction with different initial concentrations and analyzing the resultant changes in initial rates.

Methylene blue – a heterocyclic aromatic chemical compound with the molecular formula $C_{16}H_{18}N_3SCl$.

Miscible liquids – mix to form a solution (e.g., alcohol and water).

Miscibility – the ability of one liquid to mix with (dissolve in) another liquid.

Mixture – two or more different substances mingled together but not chemically combined. A sample of matter composed of two or more substances, each of which retains its identity and properties.

Moderator – a substance (e.g., deuterium, oxygen, paraffin) that slows fast neutrons upon collision.

Molality – a concentration expressed as the number of moles of solute per kilogram of solvent.

Molarity – the number of moles of solute per liter of solution.

Molar solubility – the number of moles of a solute that dissolves to produce a liter of a saturated solution.

Mole – a measurement of an amount of substance; a single mole contains approximately 6.022×10^{23} units or entities; abbreviated mol.

Molecule – a chemically bonded number of electrically neutral atoms.

Molecular equation – a chemical reaction in which formulas are written as if substances existed as molecules; only complete formulas are used.

Molecular formula – indicates the number of atoms present in a molecule of a molecular substance.

Molecular geometry – the arrangement of atoms (not lone pairs of electrons) around a central atom of a molecule or polyatomic ion.

Molecular orbital (*MO*) – resulting from the overlap and mixing of atomic orbitals on different atoms (i.e., a region where an electron can be found in a molecule, as opposed to an atom); an MO belongs to the molecule.

Molecular orbital theory – a theory of chemical bonding based upon postulated molecular orbitals.

Molecular weight – the mass of one molecule of a nonionic substance in atomic mass units.

Molecule – the smallest particle of a compound capable of stable, independent existence.

Mole fraction – the number of moles of a component in a mixture divided by the number of moles in the mixture.

Monoprotic acid – can form only one hydronium ion per molecule; may be strong or weak.

Mother nuclide – nuclide that undergoes nuclear decay.

N

Native state – refers to the occurrence of an element in an uncombined or free state in nature.

Natural radioactivity – spontaneous decomposition of an atom.

Neat – conditions with a liquid reagent or gas performed with no added solvent or co-solvent.

Nernst equation – corrects standard electrode potentials for nonstandard conditions.

Net ionic equation – results from canceling spectator ions and eliminating brackets from a total ionic equation.

Neutralization – the reaction of an acid with a base to form a salt and water; usually, the reaction of hydrogen ions with hydroxide ions to form water molecules.

Neutrino – a particle that travels at speeds close to the speed of light; created due to radioactive decay.

Neutron – a neutral unit or subatomic particle with no net charge and a mass of 1.0087 amu.

Nickel-cadmium cell (NiCd battery) – a dry cell where the anode is Cd, the cathode is NiO_2, and the electrolyte is basic.

Nitrogen cycle – the complex series of reactions by which nitrogen is slowly but continually recycled in the atmosphere, lithosphere, and hydrosphere.

Noble gases – elements of the periodic Group 0; He, Ne, Ar, Kr, Xe, Rn; known as "rare gases;" formerly called "inert gases."

Nodal plane – a region in which the probability of finding an electron is zero.

Nonbonding orbital – a molecular orbital derived only from an atomic orbital of one atom; it lends neither stability nor instability to a molecule or ion when populated with electrons.

Nonelectrolyte – a substance whose aqueous solutions do not conduct electricity.

Nonmetal – an element that is not metallic.

Nonpolar bond – a covalent bond in which electron density is symmetrically distributed.

Nuclear – of or about the atomic nucleus.

Nuclear binding energy – the energy equivalent of the mass deficiency; the energy released in forming an atom from the subatomic particles.

Nuclear fission – when a heavy nucleus splits into nuclei of intermediate masses, and protons are emitted.

Nuclear magnetic resonance spectroscopy – a technique that exploits the magnetic properties of specific nuclei; helps identify unknown compounds.

Nuclear reaction – involves a change in the composition of a nucleus and can emit or absorb a tremendous amount of energy.

Nuclear reactor – a system in which controlled nuclear fission reactions generate heat energy on a large scale, subsequently converted into electrical energy.

Nucleons – particles comprising the nucleus: protons and neutrons.

Nucleus – the tiny and dense, positively charged center of an atom containing protons and neutrons, as well as other subatomic particles; the net charge is positive.

Nuclides – refers to different atomic forms of elements; in contrast to isotopes, which refer only to different atomic forms of a single element.

Nuclide symbol – designation for an atom A/Z E, in which E is the symbol of an element, Z is its atomic number, and A is its mass number.

Number density – a measure of the concentration of countable objects (e.g., atoms, molecules) in space; the number per volume.

O

Octahedral – molecules and polyatomic ions with one atom in the center and six atoms at the corners of an octahedron.

Octane number – a number that indicates how smoothly a gasoline burns.

Octet rule – during bonding, atoms tend to reach an electron arrangement with eight electrons in the outermost shell. Many representative elements attain at least a share of eight electrons in their valence shells when they form molecular or ionic compounds; there are some limitations.

Open sextet – species with only six electrons in the highest energy level of the central element (many Lewis acids).

Orbital – may refer to an atomic orbital or a molecular orbital.

Organic chemistry – the chemistry of substances that contain carbon-hydrogen bonds.

Organic compound – substances that contain carbon.

Osmosis – when solvent molecules pass through a semi-permeable membrane from a dilute solution into a more concentrated solution.

Osmotic pressure – the hydrostatic pressure produced on the surface of a semi-permeable membrane.

Outer orbital complex – valence bond designation for a complex in which the metal ion utilizes d orbitals in the outermost (occupied) shell in hybridization.

Overlap – the interaction of orbitals on different atoms in the same region of space.

Oxidation – the addition of oxygen or the loss of electrons. An algebraic increase in the oxidation number may correspond to a loss of electrons.

Oxidation numbers – quantitative values used as mechanical aids in writing formulas and balancing equations; for single-atom ions, they correspond to the charge on the ion; more electronegative atoms are assigned negative oxidation numbers, known as *oxidation states*.

Oxidation-reduction reactions – reactions in which oxidation and reduction occur; known as *redox reactions*.

Oxide – a binary compound of oxygen.

Oxidizing agent – the substance that oxidizes another substance and is reduced.

P

Pairing – a favorable interaction of two electrons with opposite m values in the same orbital.

Pairing energy – the energy required to pair two electrons in the same orbital.

Paramagnetism – attraction toward a magnetic field, stronger than diamagnetism but still weak compared to ferromagnetism.

Partial pressure – the force exerted by one gas in a mixture of gases.

Particulate matter – fine, divided solid particles suspended in polluted air.

Pauli exclusion principle – no electrons in the same atom may have identical sets of four quantum numbers.

Percentage ionization – the percentage of the weak electrolyte that will ionize in a solution of given concentration.

Percent by mass – 100% times the actual yield divided by the theoretical yield.

Percent composition – the mass percent of each element in a compound.

Percent purity – the percent of a specified compound or element in an impure sample.

Period – the elements in a horizontal row of the periodic table.

Periodicity – regular periodic variations of properties of elements with their atomic number (and position in the periodic table).

Periodic Law – the properties of the elements are periodic functions of their atomic numbers.

Periodic table – an arrangement of elements by increasing atomic numbers, emphasizing periodicity.

Peroxide – a compound with oxygen in –1 oxidation state; metal peroxides contain the peroxide ion, O_2^{2-}.

pH – the measure of acidity (or basicity) of a solution; negative logarithm of the concentration (mol/L) of the H_3O^+ $[H^+]$ ion; scale is commonly used over a range 0 to 14.

Phase diagram – shows the equilibrium temperature-pressure relationships for different phases of a substance.

pH scale – a range from 0 to 14. If the pH of a solution is 7, it is neutral; if the pH of a solution is less than 7, it is acidic; if the pH of a solution is greater than 7, it is basic.

Permanent hardness – hardness (relative to lathering soap) in water that cannot be removed by boiling; caused by calcium sulfate.

Photoelectric effect – emission of an electron from the surface of a metal caused by impinging electromagnetic radiation of specific minimum energy; the current increases with increasing radiation intensity.

Photon – a carrier of electromagnetic radiation of all wavelengths, such as gamma rays and radio waves; known as a *quantum of light*.

Physical change – when a substance changes from one physical state to another, but no substances with different compositions are formed; physical change may involve a phase change (e.g., melting, freezing) or another physical change, such as crushing a crystal or separating one volume of liquid into different containers; it does not produce a new substance.

Plasma – a physical state of matter that exists at extremely high temperatures in which molecules are dissociated, and most atoms are ionized.

Polar bond – a covalent bond with an unsymmetrical distribution of electron density.

Polarimeter – a device used to measure optical activity.

Polarization – the buildup of a product of oxidation or reduction of an electrode, preventing further reaction.

Polydentate – refers to ligands with more than one donor atom.

Polyene – a compound that contains more than one double bond per molecule.

Polymerization – the combination of many small molecules to form large molecules.

Polymer – a large molecule consisting of chains or rings of linked monomer units, usually characterized by high melting and boiling points.

Polymorphous – refers to substances that can crystallize in more than one crystalline arrangement.

Polyprotic acid – forms two or more H_3O_+ ions per molecule; often, at least one ionization step is weak.

Positron – a nuclear particle with the mass of an electron but opposite charge (positive).

Potential difference (or *voltage*) – the force that moves the electrons around circuit; unit is Volt (V).

Potential energy (*PE*) – energy stored in a body or a system due to its position in a force field or configuration.

Power – the rate at which energy is converted from one form to another; the unit is Watts (W). Power = voltage × current (P = VI).

Precipitate – an insoluble solid formed by mixing in solution the constituent ions of a slightly soluble solution.

Precision – how close the results of multiple experimental trials are; see *accuracy*.

Pressure – force per unit area; unit is Pascal (Pa).

Primary standard – a known high degree of purity substance that undergoes one invariable reaction with the other reactant of interest.

Primary voltaic cells – voltaic cells that cannot be recharged; no further chemical reaction is possible once the reactants are consumed.

Products – chemicals produced (from reactants) in a chemical reaction.

Proton – a subatomic particle having a mass of 1.0073 amu and a charge of +1, found atom's nucleus.

Protonation – the addition of a proton (H^+) to an atom, molecule, or ion.

Pseudobinaryionic compounds – contain more than two elements but are named like binary compounds.

Q

Quanta – the minimum amount of energy emitted by radiation.

Quantum mechanics – the study of how atoms, molecules, subatomic particles, etc., behave and are structured; a mathematical method of treating particles based on quantum theory, which assumes that energy (of small particles) is not infinitely divisible.

Quantum numbers – numbers that describe the energies of electrons in atoms; derived from quantum mechanical treatment.

Quarks – elementary particles and a fundamental constituent of matter, combining to form hadrons (i.e., protons and neutrons).

R

Radiation – 1) heat transfer through invisible rays, which travel outwards from the hot object without a medium. 2) high-energy particles or rays emitted during the nuclear decay processes.

Radical – an atom or group of atoms that contains one or more unpaired electrons; usually a very reactive species.

Radioactive dating – a method of dating ancient objects by determining the ratio of mother and daughter nuclides present in an object and relating the ratio to the object's age via half-life calculations.

Radioactive tracer – a small amount of radioisotope replacing a nonradioactive isotope of the element in a compound whose path (e.g., in the body) or whose decomposition products are monitored by detection of radioactivity; known as a "radioactive label."

Radioactivity – the spontaneous disintegration of atomic nuclei.

Raoult's Law – the vapor pressure of a solvent in an ideal solution decreases as its mole fraction decreases.

Rate-determining step – the slowest step in a mechanism; determines the overall reaction rate.

Rate-law expression – an equation relating the reaction rate to the concentrations of the reactants and the specific rate of the reaction.

Rate of reaction – the change in the concentration of a reactant or product per unit time.

Reactants – substances consumed in a chemical reaction; react together in a chemical reaction.

Reaction quotient – the mass action expression under any set of conditions (not necessarily equilibrium); its magnitude relative to K determines the direction in which the reaction must occur to establish equilibrium.

Reaction ratio – the relative amounts of reactants and products involved in a reaction; may be the ratio of moles, millimoles, or masses.

Reaction stoichiometry – describes the quantitative relationships among substances participating in chemical reactions.

Reactivity series (or activity series) – an empirical, calculated, and structurally analytical progression of a series of metals, arranged by "reactivity" from highest to lowest; used to summarize information about the reactions of metals with acids and water, double displacement reactions, and the extraction of metals from ores.

Reagent – a substance (or compound) added to a system to cause a chemical reaction or to visualize if a reaction occurs; the terms reactant and reagent are often used interchangeably; however, a *reactant* is more specifically a substance consumed during a chemical reaction.

Reducing agent – a substance that reduces another substance and is itself oxidized.

Reduction – the removal of oxygen or the gaining of electrons.

Resonance – the concept in which two or more equivalent dot formulas for the same arrangement of atoms (resonance structures) are necessary to describe the bonding in a molecule or ion.

Reverse osmosis – forcing solvent molecules to flow through a semi-permeable membrane from a concentrated solution into a dilute solution by applying greater hydrostatic pressure on the concentrated side than the osmotic pressure opposing it.

Reversible reaction – processes that do not go to completion and occur in the forward and reverse direction.

S

Saline solution – a general term for NaCl (i.e., sodium chloride) in water.

Salt – when a metal replaces the hydrogen of an acid.

Salts – ionic compounds composed of anions and cations.

Salt bridge – a U-shaped tube containing an electrolyte, connects the two half-cells of a voltaic cell.

Saturated solution –no more solute will dissolve at that temperature.

***s*-block elements** – group 1 and 2 elements (alkali and alkaline metals), including hydrogen and helium.

Schrödinger equation – quantum state equation representing the behavior of an electron around an atom; describes the wave function of a physical system evolving.

Second Law of Thermodynamics – the universe tends toward a state of greater disorder in spontaneous processes.

Secondary standard – a solution that has been titrated against a primary standard; a standard solution.

Secondary voltaic cells – voltaic cells that can be recharged; original reactants can be regenerated by reversing the direction of the current flow.

Semiconductor – a substance that does not conduct electricity at low temperatures but will do so at higher temperatures.

Semi-permeable membrane – a thin partition between two solutions through which specific molecules can pass but others cannot.

Shielding effect – electrons in filled sets of *s*, *p* orbitals between the nucleus and outer shell electrons shield the outer shell electrons somewhat from the effect of protons in the nucleus; known as the "screening effect."

Sigma (σ) bonds – bonds resulting from the head-on overlap of atomic orbitals. The region of electron sharing is along and (cylindrically) symmetrical to the imaginary line connecting the bonded atoms.

Sigma orbital – molecular orbital resulting from the head-on overlap of two atomic orbitals.

Single bond – covalent bond resulting from the sharing of two electrons (one pair) between two atoms.

Sol – a suspension of solid particles in a liquid; artificial examples include sol-gels.

Solid – one of the states of matter, where the molecules are packed closely, resistance to movement/deformation, and volume change.

Solubility product constant – equilibrium constant for the dissolution of a slightly soluble compound.

Solubility product principle – the solubility product constant expression for a slightly soluble compound is the product of the concentrations of the constituent ions; each raised to the power that corresponds to the number of ions in one formula unit.

Solute – the dispersed (i.e., dissolved) phase of a solution; the solution is mixed into the solvent (e.g., NaCl in saline water).

Solution – a homogeneous mixture of multiple substances; comprised of solutes and solvents; a mixture of a solute (usually a solid) and a solvent (usually a liquid).

Solvation – the process by which solvent molecules surround and interact with solute ions or molecules.

Solvent – the dispersing medium of a solution (e.g., H_2O in saline water).

Solvolysis – the reaction of a substance with the solvent in which it is dissolved.

***s*-orbital** – a spherically symmetrical atomic orbital; one per energy level.

Specific gravity – the ratio of the density of a substance to the density of water.

Specific heat – the amount of heat required to raise the temperature of one gram of substance 1 °C.

Specific rate constant – an experimentally determined (proportionality) constant; different for different reactions, and which changes only with temperature; k in the rate-law expression: Rate = k [A] × [B].

Spectator ions – ions in a solution that do not participate in a chemical reaction.

Spectral line – any of several lines corresponding to definite wavelengths of an atomic emission or absorption spectrum; marks the energy difference between two energy levels.

Spectrochemical series – arrangement of ligands in order of increasing ligand field strength.

Spectroscopy – the study of radiation and matter, such as X-ray absorption and emission spectroscopy.

Spectrum – display of component wavelengths (colors) of electromagnetic radiation.

Speed of light – the speed at which radiation travels through a vacuum (299,792,458 m/sec).

Square planar – describes molecules and polyatomic ions with one atom in the center and four atoms at the corners of a square.

Square planar complex – relationship with metal in the center of a square plane, with ligand donor atoms at each of the four corners.

Standard conditions for temperature and pressure (STP) – used to compare experimental results (25 °C and 100.000 kPa).

Standard electrodes – half-cells in which the oxidized and reduced forms of a species are present at the unit activity (1.0 M solutions of dissolved ions, 1.0 atm partial pressure of gases, pure solids, and liquids).

Standard electrode potential – by convention, the potential (Eo) of a half-reaction as a reduction relative to the standard hydrogen electrode when species are present at unit activity.

Standard entropy – the absolute entropy of a substance in its standard state at 298 K.

Standard molar enthalpy of formation – the amount of heat absorbed in forming one mole of a substance in a specified state from its elements in their standard states.

Standard molar volume – space occupied by 1 mole of ideal gas under standard conditions; 22.4 liters.

Standard reaction – a process where the numbers of moles of reactants in the balanced equation, in their standard states, are entirely converted to the numbers of moles of products in the balanced equation, at their standard state.

State of matter – a homogeneous, macroscopic phase (e.g., gas, plasma, liquid, solid) in increasing concentration.

Stoichiometry – quantitative relationships of elements and compounds undergoing chemical changes.

Strong electrolyte – a substance that conducts electricity well in a dilute aqueous solution.

Strong field ligand – a ligand that exerts a strong crystal or ligand electrical field and generally forms low-spin complexes with metal ions when possible.

Structural isomers – compounds that contain the same number and kinds of atoms, but with different geometries.

Subatomic particles – comprise an atom (e.g., protons, neutrons, electrons).

Sublimation – the direct vaporization of a solid by heating without passing through the liquid state; a phase transition from solid to gas.

Substance – any matter, specimens with the same chemical composition and physical properties.

Substitution reaction – a reaction in which another atom or group of atoms replaces an atom or a group of atoms.

Supercooled liquids – liquids that, when cooled, apparently solidify but continue to flow very slowly under the influence of gravity.

Supercritical fluid – a substance at a temperature above its critical temperature.

Supersaturated solution – contains a higher than saturation concentration of solute; slight disturbance or seeding causes crystallization of excess solute.

Suspension – a heterogeneous mixture in which solute-like particles settle out of the solvent-like phase sometime after their introduction. A mixture of a liquid and a finely divided insoluble solid.

T

Talc – a mineral representing the Mohs Scale, composed of hydrated magnesium silicate with the chemical formula $H_2Mg_3(SiO_3)_4$ or $Mg_3Si_4O_{10}(OH)_2$.

Temperature – a measure of heat intensity (i.e., hotness or coldness of a sample); a measure of kinetic energy of an object. Units are measured in degrees, and scales include Celsius, Fahrenheit, and Kelvin.

Temporary hardness – hardness in water, removed by boiling; caused by calcium hydrogen carbonate.

Ternary acid – a ternary compound containing H, O and another element, often a nonmetal.

Ternary compound – a compound consisting of three elements; may be ionic or covalent.

Tetrahedral – a term used to describe molecules and polyatomic ions with one atom in the center and four atoms at the corners of a tetrahedron.

Theoretical yield – the maximum amount of a specified product that could be obtained from specified amounts of reactants, assuming complete consumption of the limiting reactant according to only one reaction and complete recovery of the product; see *actual yield.*

Theory – a model describing the nature of a phenomenon.

Thermal conductivity – a property of a material to conduct heat (often noted as k).

Thermal cracking – decomposition by heating a substance in the presence of a catalyst and without air.

Thermochemistry – the study of absorption/release of heat within a chemical reaction; studies heat energy associated with chemical reactions and physical transformations.

Thermodynamics – studying the effects of changing temperature, volume, or pressure (or work, heat, and energy) on a macroscopic scale.

Thermodynamic stability – when a system is in its lowest energy state with its environment (equilibrium).

Thermometer – a device that measures the average energy of a system.

Thermonuclear energy – energy from nuclear fusion reactions.

Third Law of Thermodynamics – entropy of a pure crystalline substance at absolute zero equals zero.

Titration – a procedure in which one solution is added to another solution until the chemical reaction between the two solutions is complete; the concentration of one solution is known, and that of the other is unknown. The process of adding one solution to a measured amount of another to find out exactly how much of each is required to react.

Torr – a unit to measure pressure; 1 Torr is equivalent to 133.322 Pa or 1.3158×10^{-3} atm.

Total ionic equation – the expression for a chemical reaction written to show the predominant form of species in aqueous solution or contact with water.

Transition elements (metals) – B Group elements except IIB in the periodic table; sometimes called transition elements, elements with incomplete *d* sub-shells; the *d*-block elements.

Transition state theory –reactants pass through high-energy transition states before forming products.

Transuranic element – an atomic number greater than 92; none of the transuranic elements are stable.

Triple bond – the sharing of three pairs of electrons within a covalent bond (e.g., N_2).

Triple point – where the temperature and pressure of three phases are the same; water has a unique phase diagram.

Tyndall effect – results from light scattering by colloidal particles (a mixture where one substance is dispersed evenly throughout another) or by suspended particles.

U

Uncertainty –any measurement that involves estimating any amount that cannot be precisely reproduced.

Uncertainty principle – knowing the location of a particle makes the momentum uncertain, while knowing the momentum of a particle makes the location uncertain.

Unit cell – the smallest repeating unit of a lattice.

Unit factor – statements used in converting between units.

Universal (or ideal) gas constant – proportionality constant in the ideal gas law (0.08206 L·atm/(K·mol)).

UN number – a four-digit code used to note hazardous and flammable substances.

Unsaturated hydrocarbons – hydrocarbons that contain double or triple carbon-carbon bonds.

V

Valence bond theory – proposes that covalent bonds are formed when atomic orbitals on different atoms overlap and the electrons are shared.

Valence electrons – outermost electrons of atoms; usually those involved in bonding.

Valence shell electron pair repulsion theory (VSEPR) – assumes electron pairs are arranged around the central element of a molecule or polyatomic ion with maximum separation (and minimum repulsion) among regions of high electron density.

Valency – the number of electrons an atom wants to gain, lose, or share to have a full outer shell.

Van der Waals' equation – a quantitative relationship of a state extending the ideal gas law to real gases by including two empirically determined parameters, which are specific for different gases.

Van der Waals force – one of the forces (attraction/repulsion) between molecules.

Van't Hoff factor – the ratio of moles of particles in solution to moles of solute dissolved.

Vapor – when a substance is below the critical temperature in the gas phase.

Vaporization – the phase change from liquid to gas.

Vapor pressure – the particle pressure of vapor at the surface of its parent liquid.

Viscosity – the resistance of a liquid to flow (e.g., oil has a higher viscosity than water).

Volt – one joule of work per coulomb; the unit of electrical potential transferred.

Voltage – the potential difference between two electrodes; a measure of the chemical potential for a redox reaction.

Voltaic cells – electrochemical cells in which spontaneous chemical reactions produce electricity; known as *galvanic cells*.

Voltmeter – an instrument that measures the cell potential.

Volumetric analysis – measuring the volume of a solution (of known concentration) to determine the substance's concentration within the solution; see *titration*.

W

Water equivalent – the amount of water absorbing the same heat as the calorimeter per degree of temperature increase.

Weak electrolyte – a substance that conducts electricity poorly in a dilute aqueous solution.

Weak field ligand – a ligand that exerts a weak crystal or ligand field and generally forms high-spin complexes with metals.

X

X-ray – electromagnetic radiation between gamma and UV rays.

X-ray diffraction – a method for establishing structures of crystalline solids using single-wavelength X-rays and studying the diffraction pattern.

X-ray photoelectron spectroscopy – a spectroscopic technique used to measure the composition of a material.

Y

Yield – the amount of product produced during a chemical reaction.

Z

Zone melting – remove impurities from an element by melting and slowly traveling it down an ingot (cast).

Zone refining – a method of purifying a metal bar by passing it through an induction heater; this causes impurities to move along a melted portion.

Zwitterion (formerly called a dipolar ion) – a neutral molecule with a positive and negative electrical charge; multiple positive and negative charges can be present, distinct from dipoles at different locations within that molecule; known as *inner salts*.

Frank J. Addivinola, Ph.D.

This study guide's lead author and chief editor is Dr. Frank Addivinola. With his outstanding education, laboratory research, and decades of university science teaching, Dr. Addivinola lent his expertise to oversee the development of this series.

During his extensive career, Dr. Addivinola held faculty positions at colleges and universities, including Harvard University, Johns Hopkins University, University of Maryland, and Northeastern University, and taught undergraduate and graduate-level courses in biology, biochemistry, organic chemistry, inorganic chemistry, physics, anatomy and physiology, medical terminology, nutrition, and medical ethics. He received several awards for his research and presentations.

Dr. Frank Addivinola conducted original research in developmental biology as a doctoral candidate and pre-IRTA fellow in Molecular and Cell Biology at the National Institutes of Health (NIH). His dissertation advisor was Nobel laureate Marshall W. Nirenberg, Chief of the Biochemical Genetics Laboratory at the National Heart, Lung, and Blood Institute (NHLBI). Before NIH, Dr. Addivinola researched prostate cancer in the Cell Growth and Regulation Laboratory of Dr. Arthur Pardee at the Dana Farber Cancer Institute of Harvard Medical School.

Dr. Addivinola holds an undergraduate degree in biology from Williams College. He completed his Masters at Harvard University, Masters in Biotechnology at Johns Hopkins University, and five other graduate degrees at the University of Maryland University College, Suffolk University, and Northeastern University.

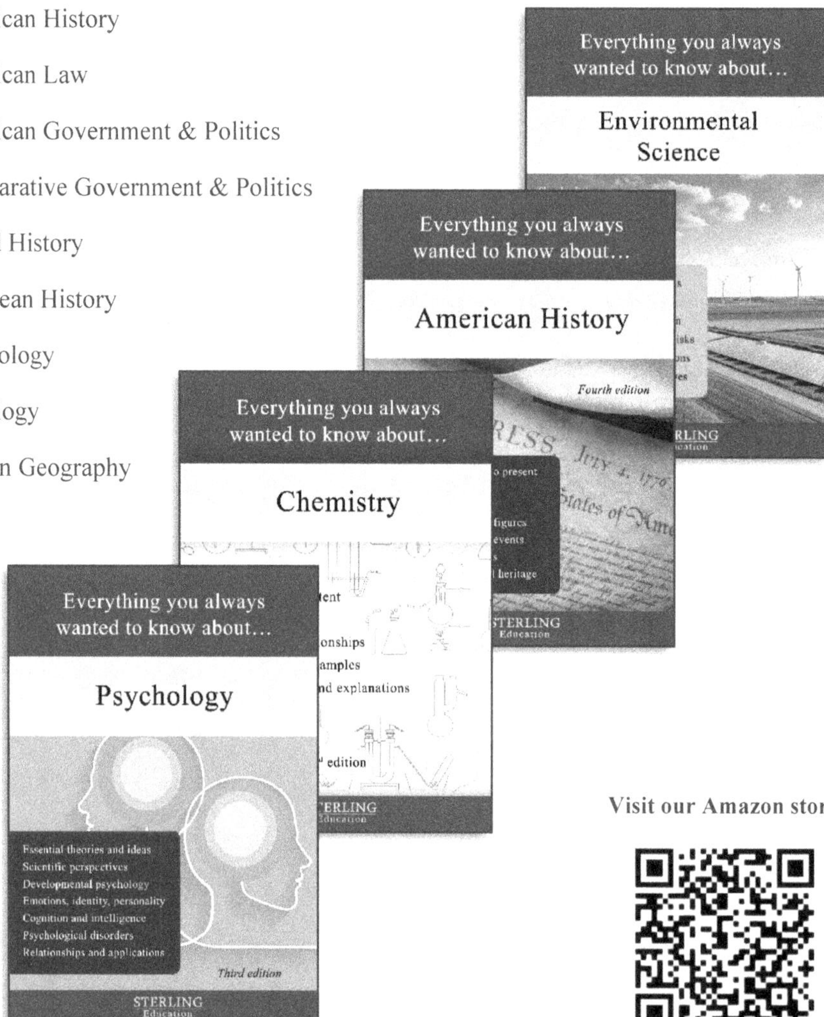

Everything you always wanted to know about…
Environmental Science
Everything you always wanted to know about…
American History
Fourth edition
Everything you always wanted to know about…
Chemistry
Everything you always wanted to know about…
Psychology
Essential theories and ideas
Scientific perspectives
Developmental psychology
Emotions, identity, personality
Cognition and intelligence
Psychological disorders
Relationships and applications
Third edition
STERLING
Education

For online practice resources visit
https://www.sterling-prep.com

www.ingramcontent.com/pod-product-compliance
Lightning Source LLC
LaVergne TN
LVHW080044210926
848481LV00011B/3277
* 9 7 9 8 8 8 5 5 7 0 6 7 1 *